Pat Treusch
Robotic Knitting

Science Studies

Pat Treusch (Dr. phil.) works as a feminist science and technology studies scholar at the Centre for Interdisciplinary Women's and Gender Studies (ZIFG) and the Department of General and Historical Educational Sciences at the Technical University Berlin.

Pat Treusch
Robotic Knitting
Re-Crafting Human-Robot Collaboration Through Careful Coboting

[transcript]

Robotic Knitting is funded by the Volkswagen Foundation and supported by MTIengAge (funded by the BMBF) and by DiGiTal (funded by the Berliner Chancengleichheitsprogramm BCP).

Bibliographic information published by the Deutsche Nationalbibliothek

The Deutsche Nationalbibliothek lists this publication in the Deutsche Nationalbibliografie; detailed bibliographic data are available in the Internet at http://dnb.d-nb.de

This work is licensed under the Creative Commons Attribution-NonCommercial-NoDerivatives 4.0 (BY-NC-ND) which means that the text may be used for non-commercial purposes, provided credit is given to the author. For details go to http://creativecommons.org/licenses/by-nc-nd/4.0/.

To create an adaptation, translation, or derivative of the original work and for commercial use, further permission is required and can be obtained by contacting rights@transcript-publishing.com

Creative Commons license terms for re-use do not apply to any content (such as graphs, figures, photos, excerpts, etc.) not original to the Open Access publication and further permission may be required from the rights holder. The obligation to research and clear permission lies solely with the party re-using the material.

© 2021 transcript Verlag, Bielefeld

Cover layout: Hagen Verleger
Cover illustration: Hagen Verleger
Proofread by: Jessica Sorensen
Printed by Majuskel Medienproduktion GmbH, Wetzlar
Print-ISBN 978-3-8376-5203-1
PDF-ISBN 978-3-8394-5203-5
https://doi.org/10.14361/9783839452035

Printed on permanent acid-free text paper.

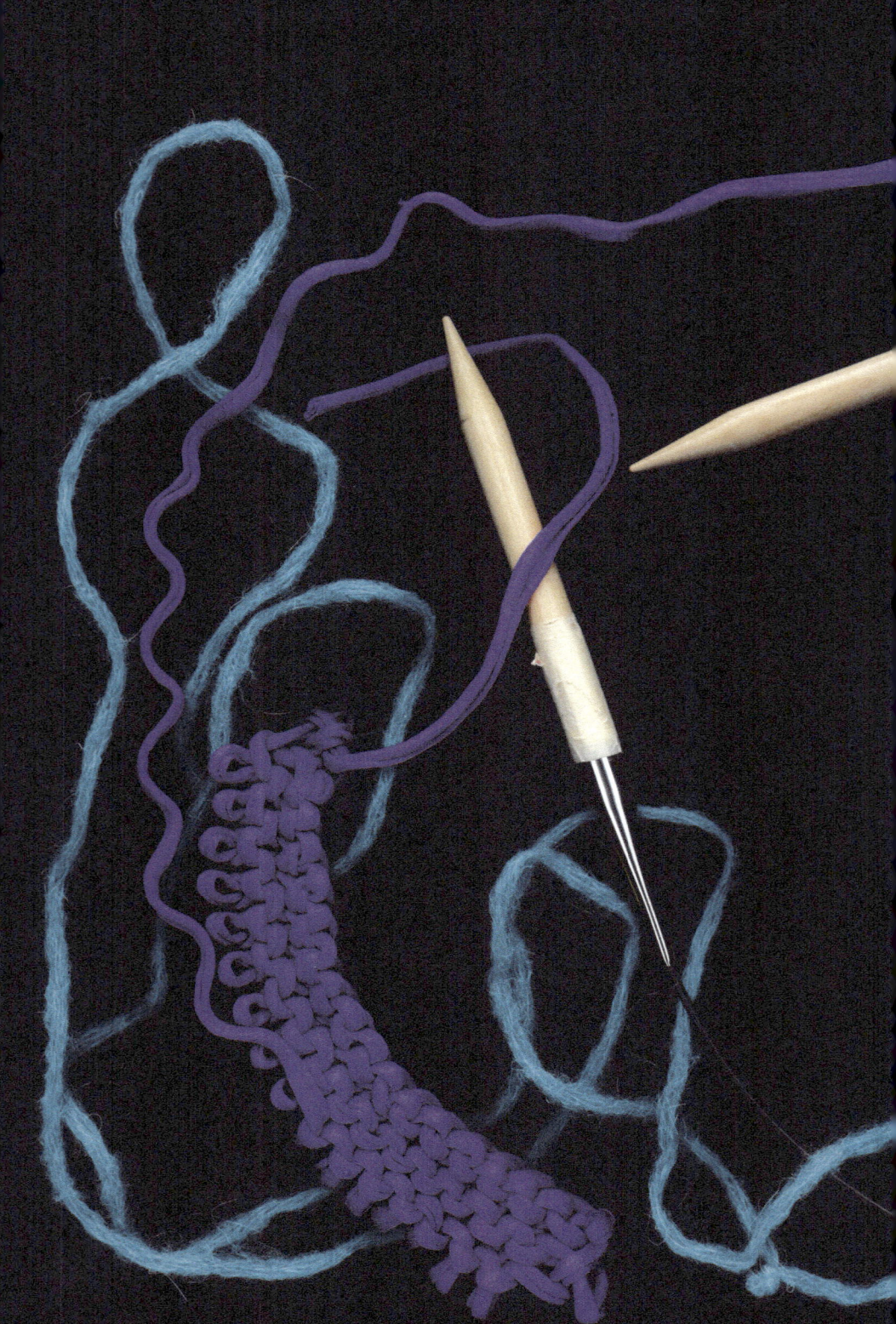

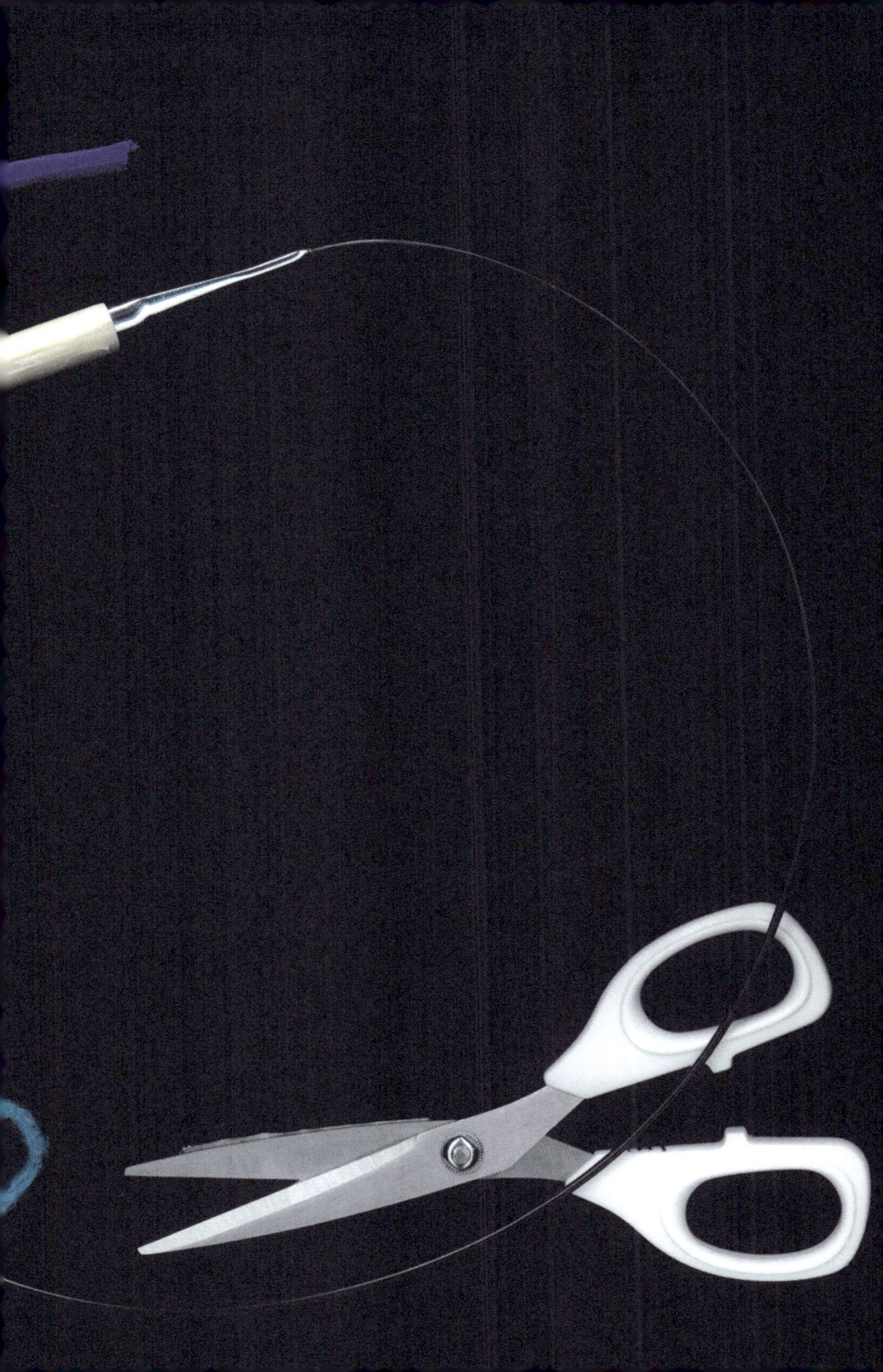

Contents

Introduction: Is the Robotic Future Open (for Knitting)?	7
Chapter 1: The Knitter in the Lab	17
1.1 Discursive Certainties? Engaging with Cobot Discourses	28
1.2 Situated Co-Engineering: An Interdisciplinary Account of Engaging with the Cobot	49
Chapter 2: String Figuring Robotic Knitting	67
2.1 String 1: Knitting and the Digital—Diffracting Dichotomous Relations	68
2.2 String 2: The Knitting Hands and the Knitting Grippers	88
Chapter 3: Knitting Together	123
3.1 String Figuring Storylines & Sociomaterial Configurations	124
3.2 Careful Coboting through Hand Knitting – and Beyond	141
Bibliography	145
List of Figures	153
Acknowledgments	155

Introduction: Is the Robotic Future Open (for Knitting)?

Our worlds of the Global North are increasingly inhabited by a number of visual and textual narratives of a robot-technologies-driven future that seems to start already now. Or, in John Urry's (2016, 1) words, "Futures are now everywhere." Here, I am thinking of popular headlines in which the arrival of the robots is announced, especially, but not exclusively, in online media. These headlines report not only that "the robots are coming," but also what they are coming for. In most cases, they depict the robotic arrival as hostile and that the robots are coming "to steal our jobs".[1] At the same time, there also exists an almost equivalent number of headlines that advertise a robotic future in which the robots are coming to serve as 'our' somewhat human-like companions that will be beneficial to 'us' and help 'us' in different social scenarios—from shopping to elderly and infant care.[2] Headlines might diverge on what the robots are coming for (hostile takeover or beneficial assistance), but what seems to be indisputable is that the robots *are* coming.

This inevitability of "the robots are coming" is at the heart of the contemporary sociotechnical imaginary[3] of robotic futures. It also carries an inescapability with it: the presence of robots as future social agents feels overwhelmingly ubiquitous and quite confusing. Contributing to this confusing

---

1 For a significant overview and collection of headlines, see *#notmyrobots* on Twitter, https://notmyrobot.home.blog. Further, to see the most recent visions, I suggest typing "the robots are coming" into the internet search engine of your choice.
2 On the figuration of "robotic companionship", see Treusch 2015.
3 I draw on the term imaginary as it has been "established in feminist studies that investigate different bio-technology-driven visions of societal futures and how these imaginings implement and negotiate understandings of subjectivity and sociality in regard to the realms of the present and the future" (Treusch 2015, 14). On *The Robotic Imaginary* see also Rhee 2018.

inescapability is the fact that it is almost impossible to get an overview of the literature written on this topic of 'our' future with robots. Nor is it easy to take stock of the different funding schemes for robotic research, the robotic initiatives, and various research locations. In addition, it is especially difficult to assemble the different opinions on what this future will look like with regard to fields of interaction, robot models, and the assumed social impact; therefore, it is equally difficult to evaluate whether 'we humans' should embrace this future, or not. The robotic future takes shape as something that is at the same time inescapable and yet rather intangible, evolving around opposing clear visions of robots as socially meaningful machines which will integrate into society, and as a threat that disrupts foundational beliefs in the role of machines as 'human tools', shaking society to its core. While the fact that "they are coming" appears crystal clear, their social impact seems to remain a topic for discussion, while the details of such an integration, especially in terms of 'our' everyday lives, remains largely unclear. Precisely this field of tensions between inescapability and intangibility, which I view as characteristic for what I frame as the contemporary robotic imaginary, seems to call for an either-or positioning towards the hegemonic picture painted of robotic futures: either a utopian, welcoming position or a dystopian, resistant position. There is a lot to lose if 'we' comply to this either-or formation—not only with regard to imagining more socially-just kinds of human-robot co-habitation, but also in transgressing some of 'our' foundational beliefs and legacies of the Global North regarding what it means to be human, in contrast to what it means to be a robot.

This book is a technofeminist intervention into the contemporary robotic imaginary and its either-or formation. It aims at situating the claim of "the robots are coming" within the debates on how robots will become socially meaningful agents as well as the concomitant practices of realising human-robot interaction (HRI). More precisely, it intervenes by establishing a collaborative practice which has not been implemented as a task of HRI so far, namely the task of collaborative knitting. Imagine walking into a robotic lab without a degree in robotics, but instead as an experienced, queerfeminist hand knitter who also happens to be a feminist science and technologies studies (FSTS) scholar, specializing in human-robot relations, bringing with you a pair of knitting needles and a ball of yarn. This setting alone might be considered an intervention as it sounds rather counterintuitive—if not fictitious—and requires an explanation. The handicraft of knitting and queerfeminist inquiry are not typically associated with the high-tech labs in which robot technologies are developed. However, looking at this setting more closely, it reveals its rootedness within the emerging field of *craft HCI (human-computer interaction)* (see Gross et al. 2013; Devendorf & Rosner 2015; Rosner 2018; Frankjaer & Dalsgaard 2018), and in my personal history of being interested in technofeminisms, robotics, and knitting. Both ultimately led to the situation described here: entering the robotic lab with a pair of knitting needles and a ball of yarn with the intent to realise human-robot knitting. Attached to this intervention, my role as a queerfeminist scholar in the robotic lab changed dramatically: from being an observer, a role with which I was already acquainted from previous research conducted at a robotic lab and on the engineering of robots as social agents (Treusch 2015), to becoming a robotics practitioner myself.

Entering the lab, I took a stance in my exploration of HRI where the goal of making a difference in debating and designing 'our' robotic future is not only to become part of the engineering of HRI, but also pivots around yarn as a material and metaphor—the red thread of this book—which enabled me to take on this role in the first place. As a knitter, I entered the lab with certain imaginations of what it would mean to make hand knitting our task of human-robot collaboration (HRC), necessarily involving an investment in the challenge of realising the handling of yarn between human and robot. At the same time, yarn in its metaphorical meaning functions as a navigational tool for exploring the contemporary sociotechnical robotic imaginary, identifying individual discursive strands in order to follow them, literally tracing that which is and that which might not yet be possible, in human-robot relations. Thus, my account of robotic knitting is grounded in my curiosity about how knitting with a robot collaboratively could challenge hegemonic narratives of the useful robot geared at helping us wherever needed *in theory and in practice*. Handling yarn then became my method to *enmesh myself* in the contemporary sociotechnical robotic imaginary on both levels: the level of discursive formations on robots as collaborative, social agents, and the level of everyday engineering practices as they take place in robotic labs. Such a becoming enmeshed is about exploring the possibilities of interrelating visions of future robots, practices of HRC, interdisciplinary knowledge, and needlework.

This book centrally builds on the experience of conducting research in, and on the resulting findings of, the interdisciplinary project *Do Robots Dream of Knitting? Re-Coding Human-Robot Collaboration (DRDK)*, funded by the Volkswagen Foundation and situated at a robotics lab at the Technical University Berlin from September 2018 until August 2019. It brings together a discussion of discourses that envision what 'our' robotic future might look like, but also an exploration of laboratory practices of enacting robotic futures through knitting collaboratively, including the experience of causing irritation precisely because of the rather unintuitive idea of making knitting a task for HRC.

I posit robotic knitting as a methodological tool and analytical frame for contemporary technofeminism. Technofeminism, in line with Cornelia Sollfrank (2018, 3), enables forms of inquiring that "mean no less than struggling for a more just and liveable world for everyone in today's technoscientific culture." Clearly, robotic presents and futures are pivotal in raising questions of more just and liveable technoscientific worlds. Getting engaged in this struggle, I took up my knitting needles and yarn to use them as the tools for producing a tangible, textile artefact together with a cobot.

Beyond the challenge of producing a knitted artefact together, to knit collaboratively with a robot also became the use case for complicating taken-for-granted certainties of 'our' contemporary sociotechnical robotic imaginary. Implementing this use case for one cobot technology, however, is not oriented at, for instance, finding the obstacles in human-machine interaction in order to make human-machine relations at this interface more efficient. Rather, robotic knitting pivots around posing these questions: How are robotic futures imagined? On what kind of human-robot relations are these visions based? How are these articulated in existing robot technologies? And, what kind of collaboration are they in turn capable of? Tackling these questions through implementing the use case of collaborative knitting, I refuse to view the narrative of "the robots are coming" as announcing an already determined course that technology development in the future will take, but rather view it as constitutive of one of the core challenges of our times: to insist on the openness of tech development. This openness also implies that the future still needs to be written. In result, I regard the process of robotic knitting as the ideal interventionist practice and generative, playful engagement with an overwhelming, inescapable, yet open situation—a position which is rooted in Donna Haraway's (1985) cyborgian sense of feminist critique.

The knowledge on human-cobot relations assembled in this book is not comprehensive in a universalizing sense, but is *partial*. Based on Haraway's (1991) *situated knowledges*, generating knowledge necessarily involves situating knowledge claims within specific arrangements of time, space, materialities, and power relations. Power relations of the contemporary sociotechnical robotic imaginary articulate, for instance, in the current hegemonic understanding of the socially meaningful robot as necessarily human-like. In her poignant analysis of the *robotic imaginary*, Jennifer Rhee (2018, 9) explains that "the metaphors we use to describe technologies are powerful actors that shape how we imagine, invent, and engage technologies and the world." In order to present the current and coming generations of robots as socially-meaningful future co-workers or workers, human-likeness has evolved into and has been established as the almost-unquestionable dominant metaphor used to describe future robotic worlds. Its legitimacy is thereby mostly based in the belief that it will guarantee that robots become socially meaningful on a large scale. Knitting collaboratively tweaks this category, as I will show in Chapter 2, by shifting the focus from human-like as the primordial category of mutual intelligibility between human and robot to the multi-dimensional practice of enacting collaboration between humans and robots.

What I am concerned with is not only the legitimacy of, or the desire for, imaginations and designs of somewhat human-like robots supposedly becoming social actors, in one way or another, and a labour source in every human sphere of capitalist production. What I am especially concerned with is the re-crafting of visions and concrete possibilities of how humans and robots can and should relate in the present and in the future, and who or what is involved in such re-craftings. According to Rhee (2018, 9), what is needed is "a more capacious vision of the robot, as well as the human". Pivotal for this book is a playful curiosity with knitting needles, yarn, and robotic artefacts as a resource for, and as agents in, developing a more capacious vision of entangled human-robot futures.

Entering a robotic lab with a pair of needles and yarn to then realise robotic knitting is a hands-on practice of intervening. Even though realising collaborative knitting between humans and a cobot appears to be a clear-cut goal, central to this project is the constant examination of the everyday practices of engineering through which we, the interdisciplinary team, are implementing this goal. Robotic knitting thus serves as a tool not only for probing taken-for-granted knowledges, but also practices of engineering such a goal, while at the same time, it also functions as a tool for re-engineering and telling a different story. This re-engineering and the story robotic knitting tells are based on bringing knitting needles, an interdisciplinary team, and a cobot together. The interventionist momentum of robotic knitting thus is at the same time disruptive and generative.

Robotic knitting's challenge of usefulness aligns with what Sara Ahmed (2019) recently terms *queer use*. Queer use is about bringing to the fore the potential of use beyond the mundane by exploring the question of: *What's the use?* Guided by raising precisely this question, my queer use of the cobot as well as of hand knitting and its materials (needles and yarn), also necessarily involved complicating not only my account of automation, the idea of robots taking over as 'our' co-workers, and the underlying organisation of work, but also my idea of hand knitting. While exploring the cobot as (co-)worker is essential to the first chapter of the book, I will show in the second chapter how choosing hand knitting as the object of digital automation tweaked my account of textile creation.

Through the DRDK project, I am not only eager to explore the ways in which the robotic future is still open, but also whether it is open to introducing knitting as a kind of *queer use case* that scrutinises use and is a valuable practice of HRC. In this sense, it is a call for a different robotic culture, much more in line with, for instance, Simone Giertz' *Shitty Robots* and her playful intervention with robots, like the lipstick robot (2016) which is filmed as smearing lipstick onto her lips, but also onto her left cheek in a very impressive fashion, while she is reading something on her tablet.[4] This robot could all too easily be dismissed as a useless invention. However, I suggest staying with the quirkiness of seemingly useless machines, such as the lipstick robot or the knitting robot, in order to re-pose the question of *What's the use?* in opening up debates on human-robot relations of the future. It is precisely in this sense that robotic knitting views the robotic future as open and works towards opening it up for knitting.

Developing the notion of robotic knitting as a multi-faceted tool, this book is split into three parts: Chapter 1 will describe central aspects of the sociotechnical formation of the robotic imaginary, as well as the methodology and methods of engagement. The second part of the book, Chapter 2, explores the relation between handicraft and cobot technologies, illustrated by the case study of robotic knitting, with a focus on practices and practicalities of realising human-robot knitting as a collaborative task. The final and closing part, Chapter 3, will present the results of this project and tentatively discuss robotic knitting beyond the case study, and as a tool for non-determinist critique: to think about and imagine 'our' robotic futures, and to act with current and future robot technologies as open and in need of a re-crafting.

---

4   https://www.youtube.com/watch?v=WcW7o-6eQcY

CHAPTER 1
THE KNITTER IN THE LAB: BECOMING SAND IN THE GEARBOX

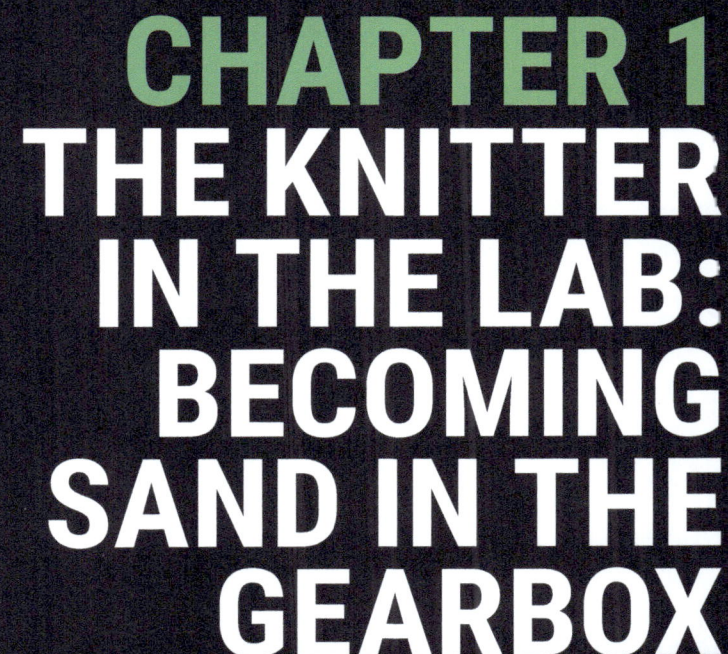

Chapter 1: The Knitter in the Lab
Becoming Sand in the Gearbox

To challenge the idea of one coherent robotic future is a technofeminist intervention into the idea of linear technological progress. I understand such an intervention in figurative terms as becoming the sand in the gears of what appears to be an overwhelmingly well-oiled machinery of 'our' robotic future. The picture of *sand in the gears* might at first glance seem like an undesirable disturbance and a very destructive endeavour. However, from my technofeminist, interventionist perspective, a second glance can reveal the productive and desirable effects of such a disturbance. In its disruptive momentum, to become the sand in the machinery of 'our' robotic future, means to take a break from technoscientific acceleration and its almost unlimited promises of improvement and optimisation of 'our' everyday lives. The sand causes a break that is involuntary and very material. At the same time, this break allows me to take up my knitting needles and yarn as tools for a queer use of the cobot.

Furthermore, becoming sand in the gear is how I imagine the workings of Haraway's figure of the cyborg—a figure that not only reclaims a techno-driven term, but also relies on such moments of involuntary break with coherent and linear stories. These moments are the motor of technoscientific worlding because they bring the potentiality to intervene into linear stories. With Nina Lykke (2010, 39), "The cyborg mobilizes other, critical stories that have the potential to undermine hegemonic power and dualisms." In this regard, I consider becoming sand in the gears as a cyborgian, technofeminist way of mobilising critical stories from within robotics. More precisely, if hegemonic narratives present collaboration as key to increasingly robotised futures, challenging human-robot collaboration (HRC) will work exactly in this way of mobilising different stories of relating and interacting between humans and machines. In addition, to fundamentally query current forms of storytelling also entails questioning the concomitant distribution of responsibility for 'our' robotic futures, as well as who is acknowledged as an expert in designing and realising liveable, sociotechnical futures. Becoming sand in the gearbox then is about reclaiming not only HRC, but also expertise in telling and enacting different stories.

Sand is also a very relevant and ubiquitous construction material with multiple fields of use. Sand can be composed of components of different sizes, encompassing gravel, pebble, and crushed rock. Indeed, sand is one of the most needed natural resources in Germany, amounting to an average use of 19 kilos per German citizen per day.[1] Quartz is one of the most important components of sand as it is not only one of the sturdiest natural materials, but also contains silicon. Its sturdiness makes it the perfect component for manufacturing glass and concrete, while the silicon has the ability to transform alternating current into direct current. With these qualities, sand is one of the foundational components for the development of microelectronics and devices like the computer chip. 'Our' techno-driven societies, and with them robotic technologies, are literally built out of sand. With regard to sand's material qualities, it is sand's sturdiness and flexibility, and other physical properties, that secure the triumph of information technologies.

In its figurative meaning, sand commonly stands for evanescence, the non-tangible, disruption, or even failure—as many popular sayings show. Pondering here about sand, my writing does not unfold linearly, but rather comes to a halt—a halt of thinking with sand. It is thinking in a different direction which orients me towards wanting to write about robots as powerful figures of seemingly linear technological progress and as solutions to societal challenges like the lack of human labour in different fields of work. Thinking with sand means to engage with sand as a basic material component of robotics, and at the same time as a potential disruptive force on two levels: First, in developing a position of critique that means to become the sand in the gears of narratives of robotic technology development. Second, in becoming aware of the materiality of robotics, including the limited availability of this natural resource. Its limitedness reminds me of the limitedness of the things 'we' build from sand—a fact that is mostly neglected in the idea of "the robots are coming". Inducing a disruptive halt appears to be an auspicious mode of engagement with productive and desirable effects for producing situated knowledge, that is, critical stories, on the contemporary robotic scene.

---

1   https://www.planet-wissen.de/technik/werkstoffe/sand/index.html

Cobot Technologies—A New Kind of Machine?

As many others have pointed out before me, the technology *robot* literally embodies the automation of work. The term robot derives from the Czech word for working, *robota*, and was coined by the brothers Karel and Josef Čapek in 1920.[2] It represents the automation of human labour. Robotic knitting pivots around the technofeminist engagement with one specific robot automation technology, namely the collaborative robot, or in short, the *cobot*. The terminology and the idea of a collaborative robot has existed for more than 20 years. The term was coined by US-American roboticists James Edward Colgate and Michael A. Peshkin in 1996, who also hold the US patent for cobots, since 1999.[3] The basic idea behind the cobot is to develop and implement robots with which new forms of collaboration—based on new forms of proximity between robots and humans—are possible. In contrast to the industrial robots successfully operating in factory halls since the second half of the 20th century, the cobot no longer has to be caged and the worker no longer protected from the large-scale, powerful robotic workforce. Rather, this new generation of collaborative robots are "interconnected, intelligent, adaptive, and are beginning to emerge from their protective cages" (Pfeiffer 2018, 21).

I suggest grappling with the cobot as a key element of the current robotic imaginary that is not only a potentially powerful labour source, but also a culturally powerful figure. When attesting that "we are in the midst of a robot invasion", David Gunkel (2018, ix) nevertheless points out that a distinction has to be made between science-fiction imaginaries of a robotic takeover and the myriad ways in which robotic technologies in different shapes and sizes have already become part of 'our' everyday tech-environments. They have historically been and continue to be important for processes of industrialisation. They are currently implemented in workplaces in industry as well as increasingly implemented in the service sector (including technologies of the so-called *Smart Home*). In this line of thought, the robotic takeover can be understood more in terms of a lingering event or a subtle set of events rather than an invasion. The picture of a subtle, successive integration of robotic technologies instead of a hostile, warlike invasion marks a shift in the narrative of technological progress as a linear, inexorable process and demands to look

---

2   In Karel Čapek's theatre play *Russum's Universal Robots (R.U.R)*, in short, artificial humans are developed in order to take over human labour as cheap work force.
3   https://patents.google.com/patent/US5952796

into mundane experiences of use, and previous decisions of introducing certain technologies into 'our' everyday lives. This shift in narration also makes necessary a shift in cultural understandings of what a (collaborative) robot is. Clearly, the idea of robots as, for instance, a tin robot moving in a very mechanical manner is not sufficient and needs to be complexified. However, the contrast model of a somewhat human-like machine, with a mostly white plastic covering and represented as an autonomous agent in Kantian terms, as pictured not only in science fiction movies but equally used in popular scientific and scientific discourses, also fails to capture the quality of contemporary robot technologies geared at human-robot collaboration.

In this regard, I consider HRC a black box and suggest opening it by understanding collaboration as a cultural *and* bodily practice that can be traced along story- and timelines, as well as through sociomaterial configurations and enactments in the robotic lab. HRC, as a complex phenomenon of new, proximate relations between humans and machines, is in need of a perspective through which a "critical examination of relevant discourses [is combined] with a respecification of material practices", as articulated by Lucy Suchman (2008, 140). This book presents such a combination of a critical reconstruction of discourses with a respecification of material practices, given through the example of the cobot. More precisely, my performative account of robotic futures brings the discourses of "the robots are coming" together with the auto-ethnographic experience of engaging with one robot technology and a technofeminist perspective on human-robot interaction (HRI). Thus, my approach to opening the black box of HRC encompasses such an opening on both the discursive level and a very literal level where I become the "affiliate human" (Suchman 2011, 119), not only in care of probing collaboration with a specific cobot model but also, and importantly, the engineering of collaboration through hand knitting.

Drawing on the legacy of feminist Science and Technology Studies (FSTS), technofeminist approaches analyse modern divides (for instance, between design and use, subject and object, and autonomy and dependency) not as given and fixed, but rather as emerging from practices of engaging, enacting, and relating (Wajcman 2004; Suchman 2007; Sollfrank 2018). As Wajcman (2004, 54) puts it: "Sociotechnical systems are not merely performed symbolically; they are also enacted materially." Technological artefacts such as the cobot are developed and brought into use through entangled sociotechnical networks of symbolic performance and material enactment. This insight opens up possibilities to explore, analyse, and transform emerging relations between cultural norms and technological artefacts through *re-configurations* of matter and meaning (Haraway 1997; Wajcman 2004; Suchman 2007). The performative framework of robotic knitting enables an intervention into the process of realising robotic futures on both the level of symbolic performance and that of material enactment of cultural and bodily practice. The focus of this chapter is on mapping the methodological and analytical approach of robotic knitting with an emphasis on the relevant debates on the symbolic order that hegemonic human-machine relations are built on and which they perpetuate.

Sociotechnical systems such as HRC are imbued with power relations and the normative operations of ordering, sorting, and constituting sociomaterial worlds. Notably, I understand neither the reproduction of a discriminatory and oppressive social order, nor the transformation of such an order, as solely either the result of human agency or provoked by technological artefacts. Rather, I am interested in exploring the interplay and multiple entanglements between humans and machines, emerging from and embedded in a historically specific set of symbolic ordering and physical conditioning, and how they either allow transgression from or perpetuate the already existing norms of what a human or a robot is, and how these can relate.

Working towards a more capacious vision of the cobot, I trace the porous and provisional nature of the performative enactment of HRI in general, and HRC in particular, as a cultural and bodily practice. Suchman (2011, 123), for instance, underlines that "the laboratory robot's life is inextricably infused with its inherited materialities and with the ongoing—or truncated—labours of its affiliated humans." Robotic knitting is interested in becoming exactly this: an affiliated human to the latest generation of cobots. This enables me to trace the truncated labours of the emerging relationship of collaboration through my own bodily experience of having to perform these labours, as I will show in Chapter 2. Thus, robotic knitting allows me to gain insights into existing technologies by experiencing what it means to become an affiliated human while blurring the boundaries between user and robotic engineer—to a certain degree. This encompasses experiencing the practices of what Benjamin Lipp (2019, 12) frames in terms of *interfacing*: "Interfacings describe the manifold processes, by which elements in various forms are rendered available for one another." The analytical reframing of the interface speaks to my approach to HRC as it emphasises the doing of interface, that is, interfacing, and attunes to the ways in which enacting human-robot interaction is based on practices of becoming available for one another. The latter encompasses forms of reciprocal relating and underlines the situational character of human-robot legibility. It is precisely such a fine-grained perspective of analysis that is needed in order to generate a more capacious version of HRC, and to become accountable as *staying with the trouble* in a Harawayan (2016) sense when implementing the collaborative task of hand knitting with a cobot. Thus, this book and the research on which it is based asks what kind of human-machine interface are collaborative robots constitutive of? What forms of interactive collaboration are they capable of? Which forms of human engagement, including the neglected, invisible labours of affiliating and interfacing, do they require? And, how and in which ways does collaboration at a given interface enable or disable certain tasks of human-robot interaction?

Becoming sand in the gear, in the manner described here, then finally aligns with Maria Puig de la Bellacasa's (2017) onto-ethico-epistemological notion of *matters of care*. De la Bellacasa (ibid., 6) works with an account of care in which "the tensions between care as maintenance doings and work, affective engagement, and ethico-political involvement...opens a terrain for exploring, in situation, the subtle thought of care, by reading these dimensions through each other." In this sense, I understand practices of affiliating and interfacing as practices of care, while practicing care in settings of HRC mainly involves *assembling neglected things* (ibid., 18), taking into account the politics of knowing in representing the actions, material entities, and affects in HRC. Furthermore, as Tania Pérez-Bustos (2017) carves out in her ethnographic research on the Columbian *Calado* embroidery: Performed by *caladoras*, needlework practices such as "unravelling and mending are constituted by care in relation to bodies and materialities" (a). Pérez-Bustos deploys an understanding of care as both constitutive for practicing needlework and constitutive for practicing an ethnography of needlework when she underlines "the knowledge dimensions...[of] craft" that are "emerging from the intimate relation between caladoras and calado materialities" (c). In this sense, I understand the practice of hand knitting with a cobot as a tool for making HRC my matter of care, including tracing neglected, invisible labours as well as cultural and bodily practices and materialities of collaboration. Acknowledging knitting as a doing and knowing, while also a very unusual scene of HRC, opens up multiple possibilities of inquiring taken-for-granted certainties and norms, and for exploring ways of relating bodies and materialities in and experiencing HRC differently.

In what follows, I will zoom in on hegemonic ideas of how humans and robots will relate in the future as a key part of the contemporary robotic imaginary. This will allow me to grapple with the overlapping storylines and cultural meanings of "the robots that are coming". Based on that, I will introduce my methodological toolbox of robotic knitting.

Hegemonic Figures: On the Symbolic Meaning of Robotic Visions

That robotic technologies historically had, have at the present, and will continue to have an impact in the future on societies is unquestioned. As Donald MacKenzie and Judy Wajcman (1999, 2) remind me: "Technology matters. It matters not just to the material condition of our lives and to our biological and physical environment—that much is obvious—but to the way we live together socially." This line of thought stipulates the following foundational question: How are robot technologies supposed to become part of 'our' societies? Becoming part of society implies becoming socially meaningful, which means becoming part of 'our' social, physical environments, but also implies pertinence to the realm of the social and forming social relations.

From a perspective of techno-determinism, there seem to exist only two ways in which the social and technology can relate: a techno-optimistic or techno-pessimistic view on a future with robots. In the first vein, the emerging robotic technologies will become useful and therefore carry with them the potential for positive change (for instance, in service sectors, including elderly care). In the second view, robots will steal 'our' jobs and therefore stand for negative effects on societies (for instance, raising—if not exploding—unemployment rates). This juxtaposition of understanding and evaluating relations between human-machine (and society-technology, respectively) in terms of either resulting in positive or negative effects, is exactly the point of departure for the longstanding technofeminist endeavour to open up, as Wajcman (2004, 6) underlines, "a way between utopian optimism and pessimistic fatalism for technofeminism, and between cultural contingency and social determinism in social theory".

Foundational for such a way beyond an either-or positioning is the rich corpus of (feminist) science and technology studies (STS) research from which technofeminism draws. Basically, technology and the social are not separate realms, but rather are conceived of as the always entangled *sociotechnical*. Then, "technology is a sociotechnical product, patterned by the conditions of its creation and use" (Wajcman 2004, 34). Every technology is characterised by *interpretative flexibility*, which highlights the role of sociotechnical relations for assigning meaning and value to a certain technology. However, and importantly, from a non-determinist perspective, sociotechnical relations have to be thought of as *reciprocal* relations of a co-constitution. Thus, Wajcman (2004, 39) concludes that "the construction of technologies is...a moving, relational process achieved in daily social interactions: entities achieve their form as a consequence of their relations with other entities." The use of the term "entity" points to the more-than-human, material entities actively involved in building and maintaining relations.

Summarizing here, my technofeminist, non-determinist account of technology means to adhere to several insights, including that technology is first and foremost a sociotechnical system, but also that: (1) sociotechnical systems are never fixed or given, but are rather processual, interactively stabilised phenomena; (2) sociotechnical processes involve humans and non-humans, persons and things; (3) these heterogeneous entities enact a technology, not only symbolically but also materially; and finally, (4) the process of enactment cannot be analysed without taking into account power relations in their normative operations, and how these are either perpetuated or transgressed.

The cobot embodies culturally powerful images of a future with robots that mainly pivots around the promise to free 'us' humans from the burdens of labour, also in spheres of labour which have been exclusively human so far—like the work of a clerk who is able to communicate in human-like ways, or even taking over as a nurse or more generally in medical care. While this might sound promising to some and terrifying to others, the potential for automating the yet non-automatable remains the same. In this regard, the cobot is envisioned to become a part of 'our' societies by stepping into multiple work relations with 'us' humans. These are necessarily *sociotechnical* relations: 'we' humans will have certain expectations of how the human-like Other will look, how it will behave, and how 'we' can engage with it. I suggest grappling with these expectations, identifications, and associations with the robot worker as foundational for the symbolic articulation and material enactment of HRC. The resulting sociomaterial configurations of HRC then emerge from a composition of social conventions, individual and collective expectations, images of robotic co-worker figures, ideas of collaboration, technological possibilities, embodiment, a range of affects such as desire, fear, frustration, but also spatial arrangements. These multifaceted ways of how technology *comes to matter* are my *matters of care*.

Against this backdrop, I will continue to explore, probe, and challenge human-machine relations of collaboration on the level of symbolic performance and material enactment as always entangled. This involves analysing the physical design, or *embodiment*, of robots, but also the resulting interactive capacities through which sociotechnical relations (that these robots should supposedly form and are forming) are made possible. Sociotechnical relations are always including human and more-than-human entities and are formed in both robotic lab settings and the realm of the envisioned use of a technology. Both realms are pervaded by power relations and concomitant cultural codes, and in both realms certain sociotechnical relations of substituting, co-working, or collaboration between humans and robots become (im-)possible over others. This forming of relations, however, is not dissolvable from bodies with capacities, expectations, and experiences, and is also mediated through visions of future scenarios of use. Hence, scenarios of use regulate and at the same time are dependent on the sociotechnical relations that can be formed between the entities involved (human and more-than-human).

Zooming in closer on the more-than-human in sociotechnical relations, I categorise cobot technologies in line with Suchman's (2011, 121) take on Haraway as "almost Human," meaning that they corporealise claims about humanness, and therefore function as *subject objects* in which the machine "Other [figures as] a differently embodied reproduction of the Self". The emerging class of current robots which are supposedly able to become agents in a socially meaningful manner is imagined and built in a way that I, 'the human', should be able to associate myself with the almost-human Other and vice versa (see Treusch 2015, 88). This should supposedly enable robots to also become social agents in private realms, that in consequence would allow new forms of automation of what used to be exclusively-human labour. Moreover, the almost Human embodies at the same time a figure of humanness as well as of the difference between 'the human' and its non-human Other. The human-like robot allows associations of humanness with the machine Other, for instance, by recognising the human Self in the humanlike Other. This, further, appears to be an essential form of relating in HRI.

Essential for a technofeminist approach, however, is to not fall into the trap of attributing a human-like agency to the robotic Other, but rather to make intelligible the more-than-human active involvement in HRI. The challenge then becomes to care for the more-than-human articulations of meaning- and matter-making beyond the pattern of the human in contrast to the human-like. How is it possible to allow for new patterns of relating to emerge? Pivotal in tackling this question is the relationality of proximity between human and robot in human-machine-interaction (Treusch 2015) that has been analysed, for instance, for its evocative (Turkle 1984), enchanting (Suchman 2007) and posthumanist, performatively enacted (Suchman 2007;

2011; Treusch 2015) qualities. These insights help me to grasp the ways in which narratives and their cultural codes, entities, affects, experiences, expectations, and space—the social and material circumstances of interaction in robotics—are factors in the realisation of HRC, as I will continue to show.

Robotic Workforce—Human Workforce—Human-like Workforce —and the Need for Clarification

This book is interested in a particular form of interaction between humans and robots—namely, collaboration. The notion *collaboration* comes from the Latin *collaborare*, to work with, and therefore implies a certain form of partnership between entities in working on reaching a certain goal or solving a specific problem together. In this regard, the idea of collaboration is tied to the idea of robots becoming somewhat complimentary to the human workforce. I argue that it is not sufficient to question how realistic or socially desired such techno-optimistic visions are. Rather, I view it as a necessity to (1) delve deeper into the complex power relations inherent to the contemporary mode of capitalist production as the globally prevalent form of production, and to (2) bring these explorations in conversation with my observations of and experience in the emergence of collaborative agency in settings of human-cobot interaction.

Key for stipulating this conversation are the processes of differentiation between 'the human' and its Others. Haraway (1991, 210) reminds me that the *universal Human* as a historically contingent figuration of power is the result of

> "the great historical constructions of gender, race, and class [that] were embedded in the organically marked bodies of woman, the colonized or enslaved, and the worker. Those inhabiting these marked bodies have been symbolically other to the fictive rational self of universal, and so unmarked, species man, a coherent subject."

What becomes crystal clear in her derivation of the figure of the Human is that processes of differentiation and hierarchizing govern and regulate not only bodily norms, but also the social affiliation of differentiated (sexed, gendered, racialised, colonialised, classed) bodies to varying societal spheres (private/public), and further, positions bodies in relation to responsibilities for and the valuation of labour. In this regard, analysing the current capitalist production and its societal labour division, in order to examine its (possible) transformation through robotic technologies, also needs to take into account the normative orderings of fields of labour, and the subjective labour force associated with this field. It cannot dissolve the universal Human in its powerful operations of differentiation and hierarchizing from capitalist production in its link to technology development. As Neda Atanasoski and Kalindi Vora (2019, 4) further highlight: "Present-day racial capitalism…posits humanity as an aspirational figuration in a relation to technological transformation, obscuring the uneven racial and gendered relations of labor, power, and social relations…of capitalist production." Asking for the ways in which cobots will shape work in the future necessarily means to ask *whose* work, including questioning which tasks are regarded as having potential for automation, who will benefit, as well as who makes these decisions.

Popular headlines mostly fail to address these complex power relations of in- and exclusion of *present-day racial capitalism*. Rather, they deploy the idea of freeing 'us humans' from the burdens of labour —whether wanted or not—and speaking to a somewhat homogeneous group of universal humans. This book works with an account of cobots that necessarily adjoins these neglected dimensions to discussions of a future with robots, and thus makes them one of its matters of care. Exploring what kind of sociotechnical relations are becoming im-/possible necessarily involves reconsidering these relations as intersectional, colonial, race, gender, class relations of the robotic history, present, and future.

Sociotechnical relations along intersectional categories of mattering are an intrinsic aspect of the power of "the robots are coming". They encompass not only the ways in which seemingly neutral robotic bodies are coded, for instance through cultural genitals (Robertson 2010, 5; Treusch 2015, 209), but also how seemingly neutral agential capacities are constitutive of specific sociotechnical relations through which culturally coded subject positions (for instance of 'the nurse') become possible at the human-robot interface.

My technofeminist intervention assesses the technological changes foreseen and envisioned to happen through robotic technologies, in order to open up ways for a renegotiation of the mattering of this technology. This then becomes a stance of critique to analyse technological change in its promise to free 'us humans' from the burdens of labour beyond determinist terms, while acknowledging the powerful differentiations between *whose* work and *what kind* of work is envisioned to be automated. Or put differently: Who is supposedly going to be freed from the burdens of what labour? How is labour redefined and revalued in robotics? Which tasks are regarded as worthy of being automated, which are not, and in which ways? And, which subject positions are made im-/possible through those design decisions? Finally, how is it possible to make a difference? This set of questions orients my analysis towards human subjectivity and agency and how lines of differentiation in their powerful operations of valuing the one over the other are renegotiated. At the same time, I situate my critique in a specific laboratory setting and a practical engagement with one cobot technology. Through robotic knitting, as I will continue to show, I immerse myself in the enactment of HCI, and therefore also in the re-crafting of a future populated by cobots.

Donna Haraway's rich work is a companion to my approach: from her foundational work on how to playfully engage with the *Cyborg* (1985) as a figure of technoscientific processes of boundary re-/drawing between human, machine, and animal, as well as between nature, society, and technology, to her more recent work on *staying with the trouble* (2016) in multi-species assemblages. One key guiding aspect of her work is how she realises a playful engagement with dreadful earthly constellations through the tools of story-retelling and re-figuring. Both take seriously the complex ways in which discursive and material mattering are entangled. Thus, the remainder of this chapter builds the ground for retelling and re-figuring robotic futures through robotic knitting.

1.1 Discursive Certainties? Engaging with Cobot Discourses

In the following subsections, I will delve deeper into the idea of a robotic workforce, how this idea takes shape in different discourses, propels new forms of automation encompassing the not-yet automated spheres of human labour, and I will present selected strands of feminist and postcolonial critiques of these visions.

Exploring different narratives and imaginations of (future) relations of humans and robots, I identify three storylines within the contemporary robotic scene. This division into three storylines is my analytical suggestion for grappling with the multi-layered dimensions of cultural meaning of the complex contemporary robotic imaginary. These storylines vary in the ways that sociotechnical relations of human-robot interaction are imagined, discussed, and critiqued. I differentiate between the storyline (A) of machines becoming workers; the storyline (B) of machines not only becoming workers, but also social agents; and the storyline (C) of machines as substituting the human workforce.

Furthermore, reading insights from these different strands together—or rather through one another—depicts a diffractive methodology of assembling the discursive formation of the contemporary robotic imaginary. Iris van der Tuin (2018, 100) writes that "diffraction is first and foremost a reading strategy that does justice to cracks in the academic canon." In the case of the picture painted on 'our' robotic future, I consider the narrative of "the robots are coming" as part of both academic and popular canons that are deeply enmeshed. Further, I understand the gap, for instance, between the announcing narrative and the robots rolling and stepping around in labs, founding the state of the art in robotic tech development, as one major crack in this enmeshed canon. The task then is to find and do justice to this and further cracks. Instead of re-narrating discursive certainties of a robotic future to come—a re-narration that literally performs frictionless automation—I, again, suggest challenging this mode of narration through disruption and deceleration. I follow the arguments of authors from different disciplines, contexts, and views that are concerned with a robotic future in one way or another, and who have sparked my interest during my project work on realising robotic knitting as a collaborative task. Further, I understand the latter in terms of making the three selected storylines players in my feminist, technoscientific recrafting of dimensions of future HRC. Haraway's take on the game of *cat's cradle* is vital for this diffractive reading—a reading in which strands become strings, and therefore introduces yet another dimension of yarn as material and metaphor to think and act with.

In 1987, *Paper Tiger Television*, a non-profit video collective, produced a video, entitled *Donna Haraway Reads 'The National Geographic' On Primates*. In the beginning of the video, she explains her approach of analysing relations of entangled nature and culture. While doing so, she holds half-unravelled balls of yarn in her hands that are messy and entangled. She then uses the yarn in her hands to describe the complexities of modern culture while she describes this as "untangl[ing] the ball of meanings" (Haraway 1987, 02:00). Untangling the yarn, pulling out strings, and thereby following one string as it leads to another that can be pulled out and followed again, is how Haraway engages with multiple layers of meaning and mattering. Notably, this is more than simply a reconstructive engagement. Rather, playing with yarn can be considered as a practice of producing new stories.

This becomes even more tangible when looking into Haraway's more recent work in which she continuously worked on the re-figurative nature of playing with yarn. Adopting the game of cat's cradle, she underlines the possibilities of generating new thoughts, but also new ways of coming to matter through string figuration. Cat's cradle is a game of producing string figures by passing loops of yarn between players. It is a practice of producing patterns with yarn. Further, she most popularly coined the game of cat's cradle as an "everyday analogy" (Lykke 2010, 155) for the technofeminist analytical tool and methodology of diffraction. Haraway (2013, 1) takes the movements of the yarn to develop a method of thinking which she describes as in the following: "Relays, cat's cradle, passing patterns back and forth, giving and receiving, patterning, holding the unasked-for pattern in one's hands." This description illuminates the ways in which yarn is a navigational tool, which one can follow, while at the same time, it is never possible to control the yarn fully and predict what will happen, how the yarn will form patterns of entangling and knotting. The yarn plays not a passive role in string figuring, but rather an active part.

As will also become tangible in the next chapter of the book, in which I will re-enter the project's robotic lab, one cannot predict the behaviour of yarn when engaging with it. Engaging with yarn thus means to take on responsibility for how stories are told, to do justice to the cracks in a canon and to acknowledge the more-than-human factors in thinking. In addition, the act of unravelling entangled yarn is also an activity that demands patience, attentiveness, and deceleration. Engaging with yarn is un-/making stories and un-/making worlds—in Haraway's (ibid.) words: "It matters what matters we use to think other matters with; it matters what stories we tell to tell other stories with; it matters what knots knot knots, what thoughts think thoughts, what ties tie ties. It matters what stories make worlds, what worlds make stories." Hence, playing with strands that become strings, forming patterns, allowing unasked-for patterns to emerge, giving and receiving, are all aspects of doing justice to the canon on cobotic futures.

Haraway's figural engagement with yarn has become central to my thinking since the very beginning of my academic work. It made me start learning how to knit during my doctoral studies and I am sure that it has also stipulated me to come up with the idea to bring a ball of yarn to the robotics laboratory where I was working when I was introduced to cobots for the first time. In this subsection, I intend to generate a more capacious vision of the human-cobot-interface. Even though built on the discursive figures available, this work exceeds a 'pure' reconstruction. Rather, I understand it in terms of identifying strings, pulling on strings, and following where they lead me. In this regard, the diffractive method of string-figuring structures my thinking and therefore also my reading and assembling of insights into the contemporary cobot discourse in what follows.

Storyline (A): Promises of Acceleration: When 'Machines Become Workers'

Reviewing current book publications on the topic of how technologies are changing and will continue to change work, Wajcman (2017) explores the status that is given to technology in narrations of future Artificial Intelligence (AI) and of robots as workforce. In line with her, a central aspect of the contemporary *futurist discourse* is the relevance of "automation, robotics and AI" (ibid., 1) in painting pictures of 'our' future. The future robot belongs to a new class of machines in these discourses. As Wajcman (ibid., 2) observes: "Machines are no longer tools; they are turning to workers themselves." Curious about how these technologies are imagined to transform society, insights into the discursive figure of the robot as worker will reveal details on the emerging relation between society and technology.

To begin with, one central dimension of how robotic technologies are imagined to change society is a shift in time as robotic workers stand for new forms of increasing efficiency through automation. Thus, the idea of a machine as worker is strongly tied to the promise of acceleration. This tie becomes tangible in popular statements like "'race with machines, instead of against them'" (ibid.). The metaphor of a race implies a relation between humans and technologies that is full of competition and the potential to lose this competition. What can be lost is open to speculation, but also indicated: If one does not engage in the race or even works against the race, one is in danger of losing one's access to (economic) prosperity, while those who engage in the race will profit. Intriguing in this is the assumption that everyone has the same chances to race with machines.

The idea of equal chances for everyone has to be read against the backdrop of the insight that hegemonic discourses tend to treat technology as a neutral and inevitable force. Challenging this suggested neutrality of AI technologies is at the core of technofeminist research and has lately become popular through work such as Safiya Noble's *Algorithms of Oppression* (2018), Kathy O'Neill's *Weapons of Math Destruction* (2016), and Mar Hicks's *Programmed In-*

equality (2017)—but also through long-standing, classical work such as Alison Adam's *Artificial Knowing: Gender and the Thinking Machine* (1998). All four books show extensively the ways in which seemingly neutral AI technologies such as the highly praised and applied methods of *Deep Learning* are shaped through the values and biases in which 'our' societies are grounded, and therefore also perpetuate sexist and racist power relations.

These works also make tangible the ways in which algorithms cannot be un-biased. Rather, the very idea of technology as neutral and free from social relations reveals itself once again as a myth. Claude Draude et al. (2019) have suggested a situated account of algorithms as a method for de-biasing, that is, becoming responsible—and through this to also acknowledge, reflect, and possibly reduce bias in algorithms. In addition, the computer scientist Anna-Katharina Zweig (2018) argues for an understanding of algorithmic decision-making as a sociotechnical system in which the social and the technical are entangled in co-constitutive relations at different stages of choosing and training algorithms, generating data, and bringing algorithms to use.

While the insight into the biased, sociotechnical nature of algorithmic decision-making appears to have been established as relevant knowledge (at least to a certain degree) across disciplinary boundaries in public and academic discourses, it can be said that this insight has *not* found any corresponding argument in the debates on the promises of new forms of robotic automation. As Wajcman (2017, 3) underlines, "political questions are too often lost in our obsession with the robotic revolution we are set to witness." In this sense, the tie between neutrality, inevitability, and robotic automation amounts to what I analyse as another crack in the canon and as the first knot in the yarn—if thinking of the canon of "the robots are coming" in terms of entangled balls of yarn. This knot is about the figure of robots as workers and in need of a careful dis-and re-entangling that consists of both untying the knot between neutrality, inevitability, acceleration, and robotic automation, and making possible new knots and patterns of robotic automation.

The robotic future deeply deploys the machine as a worker and as a neutral and inevitable force of technological change connected with the capitalist promise of acceleration. What appears to be a central dimension of the idea of the machine as worker is what Wajcman (ibid.) calls *our obsession* with emerging robots. What is the nature of this obsession? I raise this question through the perspective of making it a technofeminist matter of care. Obsession can be read as denoting a relationship towards a certain object that is characterised by a fixation on that object. Thus, I understand the use of that term here as indicating a fixation on the object 'robot as worker' that is charged with certain values. The promise of machines as workers embodies the potential of optimisation of capitalist production through both freeing 'us' humans from the burdens of labour and increasing the efficiency of human labour forces. I will return to the first dimension in storyline (C) on substituting human labour, and delve deeper into the second dimension, the increase of efficiency of human labour forces and the promises of acceleration as a sociotechnical issue, in what follows.

Again, Wajcman's rich technofeminist oeuvre is one resource for challenging the implicitness of acceleration in processes of automation by creating the seemingly universally-useful machine worker. More precisely, I am interested in her research on the relation between technology and time with her main diagnosis that "we are pressed for time" (Wajcman 2015, 4): that is, the paradox between the development of more and more supposedly time-saving technologies and the contemporary "shared experience of time poverty" (Wajcman 2018, 169). However, and importantly, the condition of being pressed for time cannot be explained through a techno-determinist perspective, but rather through sociotechnical relations of a co-shaping between technology and society. As Wajcman (ibid., 171) points out: "If we feel rushed and pressed for time, it is because of the priorities and parameters we set ourselves rather than the machines per se." One example she gives is the acceleration of communication through email. While the internet and computing capacities deliver the technological infrastructures for fast communication, Wajcman (ibid.) underlines that it is the "collective norms about appropriate response times" which have been established and which dictate an acceleration in communication combined with a constant availability. In this regard, the relation between technology and time is a sociotechnical relation that is neither determined by the technologies themselves (which guarantee absolute availability), nor by the norms about fast response times alone, but through their interplay. Further, many factors can be relevant for how this interplay takes shape: for instance, how and why certain norms are collectively accepted. These insights alone challenge the idea of a robotic acceleration in particular, but also, more generally, of the neutrality of technology and of the inevitability and the linearity of technological development.

In addition, there exists plenty of work from a historical technofeminist perspective on the introduction of household technologies, the so-called whiteware, and their promises of saving time doing chores. Classic studies are Ruth Schwartz Cowan's *More Work for Mother* (1983), Cynthia Cockburn's and Susan Ormrod's (1993) study on *Gender and Technology in the Making* that traces the development and use of the microwave, and Martina Heßler's *Mrs. Modern Woman* (2001). All three studies carve out the paradox between the promise of a relief from the burdens of chores through new technologies, and the increase of working hours despite the use of these technologies. In general, the listed studies point towards the transformation of work through the introduced technologies as a main reason for an increase in working hours: Essentially, tasks become more differentiated, refined, and specialised so that the number of tasks is growing. However, and importantly, all three studies also show how the development and the bringing to use of technologies cannot be dissolved from the power relations pertinent to the different societal spheres of production, consumption, and reproduction. From the perspective of contemporary studies on household and care work, the private is a sphere in which the distribution of and responsibility for work is divided by a global labour division along the categories of race, gender, and class. Thus, the promise of being freed from the burdens of labour in the private realm, through machines that become workers, raises multiple questions, such as: What kind of tasks are the machine workers going to take over? Whose work? And, more generally: Who is going to profit, and at what and whose cost?[4]

Raising these questions here, I underline the decontextualised nature of a promise of acceleration and the need for challenging such de-contextualised imaginations of how technologies will change societies—regardless of whether this change is believed to be for the better or for the worse. A linear acceleration of work processes depends on multiple sociotechnical factors that cannot be predicted, involving the interplay of technological artefacts, collective norms, and societal organisation of work in their regulative operations and individual everyday practices in accomplishing a task. In this way, what is needed is a more capacious understanding of the details of what 'becoming a worker' of machines implies to present and future forms of work, encompassing the societal division of labour and modes of current capitalist production, as well as the divide between production, consumption, and reproduction.

---

4   For a discussion on care in the context of the hospital and future robotic co-workers, see for instance: von Bose & Treusch 2013; 2018.

Storyline (B): When Robots Become Social & Emotional Machines

In most narratives on machines becoming workers – regardless of these workers being rather physical robots or mere virtual chatbots – the emerging workers have in common that they are portrayed as machines that will engage in proximate relations with 'us' humans. In both spheres, that of physical and that of virtual interaction, a prerequisite for such proximate relations is that "interaction between people and machines implies mutual intelligibility or shared understanding" (Suchman 2007, 34). Thus, a major concern in technology development has been the realisation of capacities for such a mutual understanding. Accordingly, a large corpus of work in the interdisciplinary field of HRI focuses on two characteristics in researching empirically and conceptualising interaction between humans and robots, namely *sociality and emotionality*. Both appear to be regarded as central premises for a successful interaction between humans and robots, and as a guarantee for increasing the mutual intelligibility between both entities.

In this subsection, I will delve deeper into the idea of robots as social and emotional machines as another crack or knot in the canon, and I ask how both aspects of mutual intelligibility are modelled onto human-machine relations. In so doing, I show how this idea connects behavioural characteristics with a corporealisation of the almost Human, with ways of finding the Self in the machine Other, and with forming bonds that are consistent with existing norms of human-human relations.

Thinking with Jutta Weber (2005, 209), I identify the emerging figure of the robotic worker as belonging to a class of technology that is characterised by a shift from "model[ling] rational-cognitive processes and...solv[ing] problems using formal structures...to socio-emotional interaction." This foregrounding of socio-emotional interaction involves both defining and modelling the social, including emotions, and based on that, designing and realising machines coherent with the established models. One early and quite famous example is the work of roboticist Cynthia Breazeal. Her key technology, the robot head Kismet, was developed in the early 2000s, with the goal that "interacting with it is like interacting with another person" (Breazeal 2002, cited in Weber 2005, 210). Here, I am not so much interested in if and how Breazeal managed to realise that goal, but rather how the robot's design was built on introducing the social and emotional robot.

When Breazeal becomes more specific about her goal, she reveals that she has a distinct form of human-human relationship in mind: that of infant-caregiver (ibid.). Suchman (2007, 237) poignantly analysed the implementation of the figure of the child as follows: "The figure of the child in Euro-American imaginaries carries with it a developmental trajectory, a becoming made up of inevitable stages and unfulfilled potentialities, that in the case of Kismet simultaneously authorizes the continuation of the project and accounts for its incompleteness." Thus, the machine's becoming social is tied to a developmental trajectory and to existing forms of relating between humans. In the case of the figure of the child, this is justified in applying the findings of developmental psychology to the design of a social robot (Suchman 2011). Notably, what is not taken into consideration are the cultural meanings of the figure of the child. This includes the care work implemented through the infant-caregiver relation and the responsibilities for this work which have been unevenly distributed individually and societally. Weber (2005) reminds us of the gendered, but also heterosexual dimensions at work (210), when she points out: "Sociality and emotionality have been deeply gendered categories in western thought that have hitherto been assigned to the feminine realm" (213). The infant robot as social robot is modelled after the symbolic ordering of the social that differentiates between a feminine and a male realm, with the effect of naturalising such binary differences when a robot is supposed to become social through female-coded modes of bonding, such as care, including physical and emotional care.

Beyond a critique of the reproduction of social relations as naturalised relations through the infant-caregiver metaphor, others have also argued that the application of certain theories of 20th century developmental psychology in robotics also serves the purpose to generate knowledge on the Human, and seeks to verify these theories through their application to machines. In this regard, Evelyn Fox-Keller speaks of a "circular trajectory" (2007, cited in Suchman 2011, 129). The danger then is that the enactment of the infant-caregiver relation between a robot and a human is not only heavily entrenched in neglected cultural norms, but also forecloses the question of what the robot's potential is with regard to the technology itself, and of what relations with humans are possible. According to Raul Hakli and Johanna Seibt (2017, 2), this foreclosing is "deeply unsettling" as it results in a "social robotics [that] is not only the engineering of robotic movements, [but also] the engineering of human social actions." This analysis of how engineering a machine that is supposed to fit into 'our' everyday lives encompasses the engineering of human social actions, first and foremost addresses how certain concepts embodied by the robot through specific design decisions are determining not only the capacities of the robot, but also what 'social interaction' with this robot means, and are therefore also defining human-human social interaction. While Hakli and Seibt underline the dimension of a potential undesired transgression of moral and ethical norms, I am—in line with technofeminists such as Susan Leigh Star (1995) and Haraway (1996)—more interested in asking: *Cui bono?* By asking who profits, I situate processes of engineering social action within specific arrangements of time, location, and power on the one hand, but I also, on the other hand, raise questions of responsibility for such an engineering of the social.

In this sense, it is important to acknowledge the individual, but also collective, dimensions of the normative character of the infant-caregiver relationality. In fact, the modern gendered, racial, and colonial labour division is key in establishing the divide between the public and the private realm, as well as between production and reproduction. As put by Sandra Harding (2008, 2), *Western modernities* are built on these foundational divides which

> "enable elite Westerners and men around the globe to escape the bonds of tradition, leaving behind for others the responsibility for the flourishing of women, children and other kin, households, and communities…. These others must do the…reproductive and 'craft' labor…. These others are mostly women and non-Western men."

Thus, the metaphor of infant-caregiver is also problematic with regard to its obscuring of the social ordering of responsibilities for reproductive and care labour, and its uneven distribution along the intersectional category of gender that privileges elite Westerners and men.

Moreover, a core component of reproductive work is emotional labour, as Jennifer Rhee (2018, 101) writes: "The robotic imaginary highlights the normative assumptions that structure emotional labor, yet another gendered and often devalued form of reproductive labor." As she further explains, emotional labour is tied to "the expression of normative emotions" which works as an "evidence for humanness" (ibid.). In this regard, the capacity to express normative emotions is held as a core capacity of humans and therefore as a requirement for robots to become intelligible, that is, social agents. The interest in emotions as a core component of humanness has been a long-standing aspect of research in AI; or as Suchman (2007, 233) underlines, "emotion is another component…needed for effective rationality." While acknowledging the value of emotions for intelligence in humans and machines, emotionality here remains nevertheless conceptualised in hierarchy to rationality and is reduced to a factor of functionality.

Furthermore, returning to emotional labours, what appears to be a dominating assumption is that the right emotions in machines will serve both to make machines more legible as social agents, but also to evoke certain behaviours in the persons engaging with such emotional machines. Kismet is one such example. As Suchman (2007) and Rhee (2018) both have carved out, the basic mode of relating at the human-Kismet interface relies on what can be understood as an activation of the person engaging with Kismet on an emotional level: The person has to not only read Kismet and in turn make themselves emotionally legible to Kismet, but also to adjust themselves to available emotional states pre-defined by this interactive setting. This host of emotional labours has to be invested in order to make the machine successful in social interaction. This also encompasses training oneself when interacting with Kismet in reading the machine and making oneself emotionally legible to it (Suchman 2007, 246). They appear to be foundational practices of interfacing and thus becoming available for one another. Working with a set of emotional labours that furthermore defines emotions through *emotional states* (Suchman 2007, 234) is based on the formalisation of emotional expression into circumscribable states that are treated as a universal quality of humanness. In addition, against the backdrop of a hierarchy between emotions, these emotional states can be considered to amount to the kind of emotions appropriate for humans, separating between appropriate and other emotions, while the latter remain not only as opposing reasoning humanity, but also as a threat to this reasoning humanity.

The seemingly paradoxical process between a valuing and devaluing of emotions in relation to rationality aligns with Sara Ahmed's (2004, 3) contemplation about the politics of emotions, when she writes: "The hierarchy between emotion and thought/reason gets displaced, of course, into a hierarchy between emotions: some emotions are 'elevated' as signs of cultivation, whilst others remain 'lower' as signs of weakness." Only a certain kind of emotions, the right kind of emotions, are tools for reasoning. In the case of AI, this functional approach to emotions is displayed in the robot Kismet's emotional setup: It is built on an established model of emotions which differentiates between six different states of emotional expression (Rhee 2018, 106). Underlying is a concept of emotions as internal states that become legible through their bodily expression. This bodily expression then has to be within one of the six states in order to be legible. Thus, the hierarchy between emotion and reason also includes the hierarchy between legible and illegible forms of expressing emotions, and is therefore normative on different levels of universalising humanness. With Ahmed (2004,4), the interrelatedness of emotionality and subjectivity secures social hierarchy while "emotions become attributes of bodies" – involving individual as much as collective bodies.

Resuming here, it is vital to dis-entangle how care as social and emotional work and expression matters, in order to re-entangle relations of mattering through care in human-robot collaboration. The expression of emotions as foundational for the robot becoming a social machine can be analysed through the powerful operations of connecting some emotions with reasoning over others, as gendered and racialised operations. A normalisation of the expression of a certain emotion over another carries with it the normalisation of a specific subject as the universal Human. The belief in a universality of emotions deployed by the idea of a mutual readability between humans and the machine, in short, is in danger of a normalisation of humanness as *Whiteness*. It is not taken into consideration, as Rhee (2018,105) points out, "how different women perform this work and what this work looks like varies significantly across gender and racial identification." In case of the infant-caregiver relation and the implied emotional labours of care as mothering, this includes the normalisation of *White motherhood*. The engineering of the social means more than just a potential transformation of ethical and moral codes of 'human sociality'. Rather, what is at stake is how emotions are categorised as either inside or outside of a certain norm of mutual readability. Those outside of this norm become illegible and therefore the individual and collective bodies to which these emotions are attributed, too, become illegible. I might even argue that emotions different to the six model emotions are dehumanised and therefore those who embody them are also dehumanised.

This book centrally aims at re-crafting the human-robot relation of interaction—more precisely, of collaboration—through first taking into account that this interface is loaded with cultural meaning, and therefore not only reproduces existing power relations, but also determines how robots and humans can relate. Based on that, re-crafting will also include to reclaim care in its emotional labours and corporeal forms as a substantial dimension of collaboration between humans and robots, and through the practice of hand knitting collaboratively, while establishing that this care is not based on the infant-caregiver metaphor, but rather carved out as a foundational practice of engineering (see Chapter 3).

Storyline (C): Robots as Replacing the Human Workforce

When machines are portrayed as becoming workers, it is not a stretch to ask what their role will be in capitalist production: If they are going to automate labour that has been exclusively human labour before, will this in consequence mean that robots will replace humans at work? Raising this question here, I want to identify and emphasise, again, two cracks in the canon of "the robots are coming": first, the universality of this claim in relation to humanness figured by *the worker*, and second, the gap between the robots rolling, jumping, or walking around in robotic laboratories and hegemonic imaginations of a robotic future that is near. Even if termed a co-worker, the robotic worker signifies a takeover at work. While I am not interested in assessing if this will happen and within what time span, I am interested in understanding the sociocultural framing of such a takeover and its concomitant redefinitions of labour, collaboration, and automation. My game of string-figuring with storylines in this subsection is geared at tracing operations of in- and exclusions inherent to the contemporary robotic imaginary around the idea of robots as replacing human workers.

Given the insights of the previous subsections, what can be immediately problematised in promises of a robotic takeover at work is that both the supposed fields of work expected to be taken over and the human labour forces expected to be replaced by robots are culturally coded as universal, that is, as the unmarked. As emphasised by Atanasoski and Vora (2019, 2, emphasis in original), "the inevitable incursion of robotics into domestic, social, military, and economic realms is commonly figured as a potential boon or threat to *all* of humanity, the figure of the human most threatened because it is iconically human...is white and male." This is foremost caused by the new forms of automation heralded by emerging robots. Fields of work and expertise have been excluded from former processes of revolutionising industrial production through automats as the non-automatable. While former industrial revolutions have mostly automated what has been classified as unskilled labour, largely performed by women and Non-Western men, the emerging robotic workforce stands for a redrawing of the boundaries of what are considered replaceable, boring tasks—a redrawing that in line with Atanasoski and Vora (ibid.) will effect *a White loss*. The promises of replicability that go hand in hand with a White loss are understood in terms of a form of liberation, which they analyse as *technoliberalism* (ibid.). Technoliberalism obscures the power relations, inequities, and hierarchical social order of racial, colonial, and gendered processes of capitalist production behind the design and development of automation technologies. It defines "what kind of tasks are replaceable, and what kind of creative capacities remain vested only in some humans" (ibid., 4). Selected tasks are considered to be replaceable while the decision behind this categorisation is presented as neutral and based on obvious evidence, rather than power relations.

One example that immediately comes to my mind is that of a receptionist robot. Existing receptionist robots are either humanoids in a white plastic covering or anthropoid robots that are endowed with robotic skin and supposedly all features of 'a human' head and torso, or sometimes even a 'complete' body in accordance with the able-bodied norm. The anthropoid body, for instance, of the robot Nadine, which was assembled in 2013, is modelled after her creator, Nadia Magnenat Thalmann (IMI Singapore).[5] One can find plenty of videos on the Internet showing either Nadine by herself or Nadine with Nadia. Nadine is supposed to figure the ideal robot receptionist: female, middle-aged, able-bodied, dressed in what can be described as formal office wear for women completed by an adequate hairdo and make-up. Watching videos of Nadine in action, the viewer can only get an impression on how the robot functions under the staged conditions of such a representative video. However, what seems intriguing is that the work of a clerk or receptionist is regarded as a replaceable task. Further, and even more intriguing, by seemingly realising these capacities in Nadine, the envisioned replacement moves from receptionist to social worker (ibid.): "She is part of the human assistive new technology which is badly needed as society cannot afford a full time social worker for each person with special needs. She can play the role of a personal, private coach always available when nobody is there." At first sight, this statement gives a strong impression that there exists a techno-fix for every societal challenge, such as integrating persons with special needs into society. This technoliberal claim does not allow one to ask, for instance, why it is regarded as not affordable to have a social worker for everyone who might need assistance. In fact, why should it be more affordable to delegate the care work and emotional labour to machines? Here again, it appears as if engineering social and emotional machines, such as the robotic assistant and social worker, are tied to an engineering of the social beyond engineering social

---

5   Citations from Nadine's webpage at IMI Singapore: https://imi.ntu.edu.sg/IMIResearch/ResearchAreas/Pages/NadineSocialRobot.aspx

action. Rather, social engineering involves both defining a problem in society (like the lack of institutionalised social work and a resulting lack in social workers) and delivering a possible tech-solution. At the same time, the envisioned assistive robot is also described as a personal, private coach that helps against social isolation as *she* will be "always available when nobody is there" (ibid.). Social isolation appears to be another issue that is in need of social engineering through techno-solutionism, while the position of the personal coach furthermore combines a spectrum of roles and functions: This can be the role of a here female-coded assistant in a difficult situation in life, or the role of a guide on reaching a certain goal, mostly geared towards an optimisation in one's personal life. In addition, the receptionist as well as the social worker are both job functions that are predominantly pursued by women and that require different forms of training, from an apprenticeship to college education. Modelled after her creator, Nadine becomes the representation of the White, college-educated, middle-class woman, and at the same time the representation of the fields of work this kind of robot will take over (namely those of White, college-educated, middle-class women, supposedly including social work, but also the whole range of the education system). She might be regarded as emblematic for a (gendered) White loss.

Debating Dehumanisation: Robots as Workers with Rights or Robots as Slaves?

While robots seem to advance into machines that are not restricted to the category of so-called 'unskilled' work, they should become, at best, useful machines to 'us humans'. Staging the useful machine as an auspicious technology *because* of its potential to replace human labour force, this narrative requires the figures of thought to tell the story of both the potential and the harmlessness of this technology. These figurations also regulate 'our' relationship with the future robots. There exists a long-standing tradition in robotics to draw on the metaphor of master-slave (see Weber 2014; Gunkel 2018), which in line with Weber (2014, 192) "describes a control relation between the expert and the machine." While Weber (ibid.) attests that the word robot etymologically contains the notion slave through its root in the Czech *robotnik*, she also argues that this metaphor loses its relevance as soon as the robots are stepping out of their formerly restricted contexts of industrial work into unrestricted settings of human everyday lives, as the latter context no longer relies on machine control as the guiding principle of relating between human and machine, but rather on interaction. However, reading through contemporary, mostly philosophical, literature on human-robot interaction, the question of control, and with it that of abuse of power, over the machine Other becomes of new relevance.

This relevance is articulated in what Gunkel (2017, 9) frames as *Instrumentalism 2.0*, which takes the advances in robotics figured for instance in machines like Nadine, but also constantly announced through the narrative of "the robots are coming", to argue for a human-robot relationality in which robots remain the "mere tools of human action," regardless of "how sophisticated they become." Thus, the robot is a technology that is constitutive of what Gunkel (ibid.) calls "a new class of instrumental servant or slave." What intrigues me is the seemingly unquestioned equation of instrument and slave in the canon on HRI and the de-thematising of the dehumanising quality implied in this equation. This also becomes very tangible in Gunkel's interchangeable use of the notions *Instrumentalism 2.0* and *Slavery 2.0*.

However, Gunkel is not an agent of deploying the perspective of instrumentalism or Slavery 2.0. Instead, his concern with this wording is that it might no longer be an appropriate choice when robots become social. His primary argument for a cessation in using the slave metaphor is not the violent and still harmful colonial legacy of this term, but rather the limits of what is thinkable if the very debate on robot rights is dismissed as not worth having (see Gunkel 2018). Furthermore, this leads to Gunkel (ibid.) to becoming a kind of advocate of *robot rights*. As he explains, the issue of robot rights mainly encompasses two dimensions: the dimension of *ontological capacities* and that of *normative obligations* (ibid., 20). With this, Gunkel raises the important issue of first asking if robots can become social agents with moral subjectivity, and second if 'we' humans think that this is a desirable status, taking into consideration the societal consequences this will have. As he further underlines, his goal is not "a simple 'yes'/'no' response" (ibid., 40) to this question. Rather, Gunkel (ibid., 42) is interested in the "critical task...to identify, explicate, and evaluate the oftentimes implicit operating systems that makes the discussion and debate about robots and rights possible in the first place." Thus, he is interested in re-envisioning categories of thought with regard to sociotechnical futures, and a special attention towards the denial and dispossession of rights as they have been essential to Western modernities.

In diagnosing that robot rights currently account for *the unthinkable*, Gunkel (ibid., 99) is working towards an expansion of the imaginable limits of envisioning how robots will matter as social agents and workers in the present and in the future. However, reading through Gunkel's *Robot Rights* (2018), one can acquire a quite comprehensive impression of how common the use of the terminology slave is, while the dehumanising effects deployed by the appellative practice of equating robotic instruments and slaves remains unnamed. Notably, this practice is far from innocent. Rather, drawing the boundary between machines as slaves and machines as social agents almost equal to the universal Human is in danger of ignoring the deep connections between dehumanisation and labour, as well as of reinforcing historical racial oppression—also in its histories of emancipatory resistance.[6]

In this regard, the very vision of robots as replacing the human labour force cannot be dissolved from asking, again: *Cui Bono?*—Who will profit from a replacement of human workers thought of and realised as an enslavement of the machine? Whose work is regarded as replaceable by whom or what? And, whose work will remain undervalued and not worthy of automation with regard to a continuation of current modes of racialised and gendered capitalist production? These questions point toward processes of boundary-drawing between the Human and the dehumanised Other through the societal reorganisation of the establishment of robotic labour forces. What is feared as either an impending White loss or as the lost chance to acknowledge the emerging robots in their capacities for humanlike labour and for the same rights as human workers, is a discursive configuration that fails to address the intimate tie between dehumanisation and labour in its colonial and sexist histories. This failure can be regarded as a crack in the canon.

Disentangling the tie between dehumanisation and labour further, I briefly look into strands that (1) connect the emergence of robots as workers with the question of *human welfare* (Birhane & Dijk 2020), and (2) develop an account of the *surrogate human effect* as "the racial 'grammar' of technoliberalism" (Atanasoski & Vora 2019, 5). Reading these strands as strings that become partners in my game of cat's cradle with the figure of the robotic worker, I regard my feelings of unsettlement with hegemonic debates on a replacement of human workers through the enslavement of robots as guiding this selection of strings.

Abeba Birhane and Jelle van Dijk (2020, 1) argue for a shift in debate – from asking if 'we' should grant robots rights or not, to making human welfare the primary concern. Mainly, they argue that

---

[6] It also ignores current feminist, afro-futurist re-workings of the relation between slavery and robotics (see especially the work of Janelle Monáe in Rhee 2018, 207).

"robot rights signal something more serious about AI technology, namely, that, grounded in their materialist techno-optimism, scientists and technologists are so preoccupied with the possible future of an imaginary machine, that they forget the very real, negative impact their intermediary creatures—the actual AI systems we have today—have on actual human beings." (ibid., 2)

Reducing the analytical perspective on the future robotic worker and their rights, first, relies on a belief in techno-optimist fantasies of creating a machine that will be human-like—at least human-like enough to grant rights similar to 'us' humans, and second, de-thematises the sociotechnical effects of already established technologies for individuals and collectives.

Birhane and Dijk (ibid., 2) offer a perspective in debating robots as a replacement of the human labour force that is not blinded by techno-optimist visions of the capable machine, which they frame as *techno-arrogance*, but instead articulates through an "ethical stance on human being...[through which] being human means to interact with our surroundings in a respectful and just way." Furthermore, they regard the role of technology as that of promoting exactly such a stance. Technology then becomes a mediator, in a phenomenological sense of a "lived embodied experience, which itself is embedded in social practices" (ibid.). This then leads to a shift in conceptualising the human-machine relationality, in short, along phenomenological insights into the embodied, distributed nature of sociotechnical interaction.

Focusing on the reduction of debates on robotic futures through the enslavement or rights divide, I agree with positions such as Gunkel's that plead against such a reduction and for a critical revision of how the relations between humans and these robots have been and still are debated. However, I am less concerned with arguing for the liberation of the one over the other or vice-versa. The master-slave-metaphor, as so commonly deployed in this debate, seems to allow only one view, namely precisely that of pinning the liberation of the one against the oppression of the other. Either the universal Human is liberated from labour at the costs of robots as mere tools, or robots become fully acknowledged as (potential) social agents with rights which means that the universal Human has to give them a place in the as yet exclusively human-confined sphere of the social. Such a place brings with it both the privilege of the robots' protection against destruction by humans, but also the potential to become a threat to White Human privileges. It matters how 'we' imagine robots as becoming a workforce equal to human labour forces.

Birhane and Dijk argue in a similar vein when dismissing the enslavement/rights debate. They differentiate between varying intentions in using the term *slave* for future robotic workers. They draw, for instance, on the plea to understand robots as slaves exactly for its dehumanising qualities (ibid., 1). In this line of view, the dehumanising logic of the slave as a social status is willingly employed to draw a boundary between 'us' humans and the machines. For Birhane and Dijk (ibid.), however, this practice of classification is built on a false logic when they explain that "one has already implicitly 'humanized' the robot, before subsequently enslaving it." In their reading, enslavement is problematised not only because of its dehumanising logics, but the implicit humanisation, which in the case of robotic workers, means to fall into the trap of techno-optimistic fantasies. Moreover, they emphasise that "By putting actual slaves, women, and 'other races' in one list with robots, one does not humanize them all, one dehumanizes the actual humans in the list" (2020, 3). What appears important to bear in mind is that the debate on the status of robots as either slaves or humanlike workers with rights seems to lead to what I identify as an impasse. The operations of enslavement certainly are foundationally based on the violent, radical challenging and even neglect of humanness in the enslaved. Acknowledging the dehumanising logics of enslavement, however, cannot be dissolved from rebellion and the emancipatory liberation from enslavement. Cutting them apart by insisting on the fact that robots will always remain non-human machines, is in danger of erasing histories of enslavement and emancipation from individual and collective memory. This then perpetuates racial and sexual oppression in the name of liberating robots when 'we' humans would, in consequence, become *slave owners* (ibid., 4). Nevertheless, a call for an emancipation and liberation of robots in the name of racialised Others and women is not a solution here, as it buys into the logics of dehumanisation with its relational ordering of formerly categorised groups of humans, an ordering that serves the purpose to secure the privileges of some at the costs of others. What seems to be at stake here is to overcome both the master-slave relation in robotics as much as the idea of robot rights in their implicit and explicit logics of Othering and dehumanisation.

Finally, I briefly turn to what Atanasoski and Vora (2019) term *the surrogate human effect*. In short, following their arguments, the task becomes to overthrow the whole idea of robots as workers in the first place. This is because they analyse the idea of replacing human workers through robotic workers, which are idealised as the more-efficient, stronger, and faster labour force, as a form of surrogacy that is foundational for Western modernities and deeply entrenched in racial and sexist operations of power. Replacement through automation technologies thus exceeds its meaning as a simple substitution of labour forces. Rather, it plays a key role in sociocultural processes of differentiating, classifying, and ordering humans with the effect of producing the figure of *the liberal subject* (ibid., 5). The liberated status of this subject depends on—in line with Atanasoski and Vora—"the racial unfreedom of the surrogate" (ibid.). In this regard, they underline how the very idea of substituting depends on processes of dehumanisation through which the Human as a concise figure emerges. Notably, Atanasoski and Vora (ibid., 7, emphasis in original) deliver a very rich analysis of "the social impact of design and engineering practices intended to replace human bodies and functions with machines *and* the shift in the definition of productivity, efficiency, value, and 'the racial' that these technologies demand in their relation to the post-Enlightenment figure of the human." I regard their work as a vital and important contribution to and necessary complement to feminist concerns with transformations of the sociomaterial grounds of agency and lived experience, as well as with transformations of the symbolic through changing relations across human and machine from a postcolonial perspective. Again, the idea of replacement is far from innocent, and the technologies developed to take over (such as the robotic worker) are not neutral artefacts, but are rather political sociotechnical agents. Their design, development, and use are processes that stipulate the production of difference between humans, but also between humans and non-humans; and at the same time, these practices are pertinent to existing forms of differentiation. To grapple with *technoliberalism*, in its logics of liberation and oppression, requires such a comprehensive account of current and historical powerful operations of differentiating and hierarchical ordering in order to open up possibilities for more just visions of human-machine relations. With Atanasoski and Vora (ibid., 8), what is needed are "projects focused on creating technologies that blur the boundaries between subject and object, the productive and unproductive, and value and valuelessness, thereby advancing structures of relation that are unimaginable in the present." I consider robotic knitting such a project, precisely concerned with advancing structures of relation that are unimaginable in the present, as I will continue to show.

Notably, while Atanasoski and Vora are also working on the imaginable limits of the present, their work presents a different approach to the unthinkable: Their engagement with the unthinkable is not about deciding whether or not the robots that supposedly will become advanced enough machines to replace 'us' humans at work should have the social status of slaves or be granted the same rights as the liberal subject. Rather, as they show, it is about dismissing the idea of a replacement from the start, and with it the very operations of ordering in which those differentiations are grounded. Instead, debating robotic futures then becomes a point of departure to open up possibilities for imagining a redistribution of responsibilities for, and the burdens of, labour beyond the *racial grammar* of the *surrogate human effect* of capitalist production.

Moreover, drawing on Haraway's string-figuring as a method for rereading insights through one another, to ignore the tie between dehumanisation and labour, in all of its facets, as it is foundationally implicated in the figure of the robotic worker, performs a cut through an existing knot —instead of engaging in the work of carefully disentangling strings in order to trace how race, slavery, robots, and labour are knotted together. The consequences of such a cut are not only epistemological, but also ontological. Re-crafting robotic futures necessarily means to explore how humans and machines can, could, and even should relate, and how responsibilities for labour of various kinds should be distributed among newly emerging robot-human interfaces. A narrowing of this relation to one of control over mere tools that assigns all responsibility to 'the human' appears to be too reductive in light of emerging robot technologies, but also with regard to a theorising of the sociotechnical in its co-shaping relations. What is at stake here is to imagine and to create new possibilities of relating that move beyond the humanlike partner/dehumanised slave divide. As I will continue to argue, this issue necessarily needs to be tackled through interdisciplinary, empirical, qualitative research that nevertheless takes into account long-standing power relations, inequalities, and bias. Working towards an erosion of the *surrogate human effect* of technoliberalism, I ask: How can the inevitable connection be dismantled between future robots becoming social agents and becoming workers that take over 'our' dull and dirty work? Robotic knitting tackles this question through a performative stance that combines the implementation of the practice of knitting with a cobot with delivering conceptual impulses for a re-crafting. A description of my methodological approach is at the centre of the next subchapter, while Chapter 2 will present the practice of robotic knitting. More precisely, my game of cat's cradle will interchange between enacting a diffractive methodology with yarn as a metaphor and a material of re-crafting in theory and in practice, blurring the boundaries between both. In this way, I take up wool as literally the material for creating new patterns of thinking and enacting human-robot relationality.

1.2 Situated Co-Engineering: An Interdisciplinary Account of Engaging with the Cobot

One core challenge of robotic knitting was to establish an interdisciplinary practice of realising collaborative knitting between humans and a cobot. While interdisciplinarity has advanced into a kind of buzzword for research, especially for research which addresses fundamental societal issues such as the future of (the automation of) work, what kind of research practice and knowledge exchange interdisciplinarity can instigate, may nevertheless differ a lot. In what follows, I will first discuss why and how robotics is a highly interdisciplinary field, identify some boundaries of interdisciplinarity as (not only) I have experienced them so far, then present my suggestion for an account of situated co-engineering, as evolving through the project work of *Do Robots Dream of Knitting? (DRDK)*.

Robotics can look at a long-standing plea for and practice of interdisciplinarity, which has been deployed by roboticists such as Rodney Brooks in the early 1990s. In short, Brooks (1991) argued for a bottom-up, instead of a top-down, account of intelligent behavior in humans and in machines that centrally builds on the relations between perception and action, as well as bodies and environments, advancing into the *embodied and embedded AI* paradigm. He also flags the importance of a collaboration between diverging fields of knowledge, for instance, when reviewing biological fields of knowledge (especially neuroscience, ethnology, and psychology) as delivering important insights into the very nature of cognitive behavior in humans and other animals that robotics should take into consideration. Brooks (1991, 23) closes his influential paper *Intelligence Without Reason* by asserting that the project of AI "is a complex endeavor and we sometimes need to step back and question why we are proceeding in the direction we are going, and look around for other promising directions." For Brooks, biological fields of knowledge are such other promising directions. Notably, what he underlines is the need to question and to look around as forms of reflecting on both disciplinary knowledge and the established canon, while he does not formalise this process or the relationship between the cooperating disciplines.

Since Brooks' new foundation of AI, robotics has advanced into a field of research and engineering which has become increasingly important, not only for developing automation technologies in the sphere of industrial production, but rather also for every sphere of human everyday lives, as argued throughout this book. Even though most robots are not yet stepping or rolling out of their protective cages in factory halls, or their confined lab environments, the popular discourse as well as the scientific endeavor are pivoting around such visions of "the robots are coming". In 2007, the roboticist Stefan Schaal announced *The New Robotics – Towards Human-Centered Machines*. Schaal (ibid., 1) defines such new robotics as targeted at the realisation of "more human-like robots [that] can live among us and take over tasks where our current society has shortcomings." The everyday robotic interface, however, brings a range of challenges with it when the central technology, the human-like robot, supposedly should be operated in everyday life and should be operable by possibly every person, regardless of their technical affinity, age, educational background, and abilities. This, in Schaal's (ibid., 3) view, should be reached by a new robotics as "require[-ing] a new kind of scientist that can traverse a very broad range of different disciplines." These different disciplines encompass—besides the classical discipline of mathematics—also biology, neuroscience, psychology, and ethics. All disciplines named here are potentially regarded as a resource for improving the engineering of human-like robots as social agents. Mentioned, for instance, is the acceptance of a robot among two social groups, namely the elderly and children. The suggested solution is—in line with Schaal (ibid., 2)—that the robot "needs to adhere to certain social behaviors and standards that we as humans find acceptable." This, then, is also where psychology and ethics are supposed to come into play.

What reads at first sight as a meaningful division of tasks along a complex goal, namely, to create human-like robots useful to humans in their everyday lives (more specifically, to the elderly and children), also operates with certain unquestioned assumptions, mostly around hegemonic beliefs in the necessity of human-like, anthropomorphic machines, in order to realise human-centeredness and user-friendliness. In the preceding sections, I already problematised this belief with regards to its perpetuation of power relations along the categories of race, gender, and class, given the example of the human-like co-worker or worker robot. My argument here is that the ideal of the anthropomorphic robot in its status as a taken-for-granted certainty and scientific foundation of (new) robotics equally needs to be questioned and revised. It cannot be factored out from the reflective process of knowledge practices as stipulated by Brooks and continued by many others, among them Schaal's proposal for a renewal of robotics.

The hegemonic belief in human-likeness as the foundational, unquestionable design scheme for realising collaborative, human-centred, and user-friendly robots is a core example for what I identify as current, but nevertheless longstanding, boundaries of interdisciplinary research. As many (feminist) humanities scholars have pointed out, this belief follows a circular logic: Katherine Hayles (2005), for instance, shows that such a belief in human-likeness can be analysed as a two-cycled co-shaping phenomenon at work. Hayles (ibid., 132) describes these mechanisms as the "use [of] a rhetoric that first takes human behavior as the inspiration for machine design and then, in a reverse feedback loop, reinterprets human behavior in light of the machines". The reverse feedback loop in its two-cycled movements makes tangible the circular logic between arguing for a human-like shape of the robot as a seemingly intuitive design decision and the idea of engineering social action: for instance, defining social behaviours and standards for homogenous conceptions of groups like 'the elderly' that the robot, in turn, will embody and that the users will have to adhere to. Thus, to leave the design ideal of human-likeness out of the realm of questionable knowledge and practices in robotics privileges the design decisions based in computer science and engineering over investigating the social, and more specifically, social practice in all its diverse shapes. In this regard, I argue, first, for including critical, social research of diverse, lively, sociomaterial worlds into the list of resources for reflecting on processes of technology development in AI in general, and more specifically in robotics. Second, I argue for re-considering the relationship between humanities, computer science and engineering, with regard to interdisciplinary collaborations.

To reflect on the scientific foundations and knowledge practices of robotics necessarily also has to involve taking into account existing power relations on both levels: the level of knowledge politics and the level of sociomaterial everyday lives, as they both condition possibilities for (inter)action. Feminist science and technology studies (FSTS) have argued for this for more than 30 years, if beginning with Haraway's influential *Cyborg Manifesto* (1985). It is indeed the emancipatory, transformative potential of FSTS in its call for a critical reflection of knowledge and engineering practices that I regard as the grounds for interdisciplinary research across disciplinary boundaries and barriers.

From my experience, there exist plenty of barriers in interdisciplinary project work, deeply rooted in disciplinary cultures. One that I have been confronted with in the past is that humanistic perspectives too often appear to be understood as just another tool for re-assuring that there exists a techno-fix for every issue or problem that might be raised through the process of tech development: from de-biasing algorithms, making robots social agents, to data security concerns. Jutta Weber (2010, 12), for instance, analyses the increasing evolvement of interdisciplinary approaches within the last two decades as part of a transformation of scientific cultures towards the implementation of a *technorationality*, characterised by a reductionist tendency and geared at ever new technoscientific solutions. In contrast, Gender Studies, as an inherently interdisciplinary field, is built on implementing the tools and methodologies for critically reflecting disciplinary certainties and taken-for-granted knowledge (see Lykke 2010). However, there also seems to exist a gap between what could be analysed as imaginations and expectations of what humanistic inquiry is and does, and what I, for instance, as an FSTS scholar with a certain specialisation of research, can contribute to HRC. I identify with this a gap between problem-solving and problem-raising cultures of research that divide between useful and useless perspectives in reaching a certain goal. Further, from the long-standing feminist concern with the politics of knowing, I also identify politics which assign value to some perspectives over others. The result of such a valuing, in short, is a hierarchisation of views and disciplines with regard to their relevance to AI research. An essential part of my research is to work against such a hierarchisation, exploring how to overcome such barriers and to engage with each other's work in a manner which leads to new, interdisciplinary insights, based on the radical, critical reflection on scientific foundations and certainties. The remainder of this chapter will present how robotic knitting not only invites such interdisciplinary research in theory and practice, but also how it became a prerequisite for realising this technofeminist intervention into hegemonic imaginations of a future with cobots.

Entering the Lab to Play with Wool & Knitting Needles

The idea to engage with a cobot through knitting grew out of a situation in which I was already working at the robotic lab of the BMBF Nachwuchsforschergruppe MTI-engAge at TU Berlin as a parental leave cover. During that time, I was asked by the head of the lab to conduct qualitative human-robot interaction studies with the robot *Pepper*. More specifically, the task was to generate knowledge on what we called the *grasping rationale* in scenarios of HRI in which the robot hands over an item to humans. A core question was: How can this almost always naturally occurring motion between two (universalised) humans be formalised and transferred onto the robot? Of course, we also defined a specific case and focused on the concomitant motion: The robot was imagined to hand over a ball while our interest was to understand when a person can rationalise which arm to use in order to take the ball from the robot. Basically, the idea was to improve the coordination between a robot's programmed intention and the humans in interaction, and thus mutual legibility. Of course, this setting operates with a set of pre-assumptions about the nature of human-human interaction and the model quality of these for HRI. Thus, conducting the experiments, my role was at the same time to deliver insights into the grasping rationale given our specific setting of HRI and to attest, and evaluate the conditions for and practices of enacting successful interaction at the human-robot interface. One salient practice was how each individual tried to realise mutual legibility between the robot and themselves with varying degrees of success (see Graf & Treusch 2019). Such a perspectivation works with Suchman's (2007, 69) ongoing work on "situated action", which "underscores the view that every course of action depends in essential ways on its material and social circumstances." As our experiments with *Pepper* have also shown, interaction between humans and the humanlike Other is enacted in such HRI settings. Key for understanding and developing the interface are the sociomaterial conditions and practices of enactment.

Furthermore, robotic knitting also tweaks the hegemonic understanding of agency as it is deployed by the category of the human-like, in comparison to the human. Petra Gemeinboeck and Rob Saunders (2016, 159) carve out that "robots play an important role in probing, questioning and daring our relationships with machines". They further suggest "[looking] at a machine's agency through the lens of performance" (ibid., 159-160). Such a lens enables a probe of relations *in practice* while taking into account the cultural codes relevant to a device's social meaningfulness and the performative quality of a machine's agency. Both are not simply characteristics and abilities a device is endowed with, but rather they emerge from its context of interaction. With Gemeinboeck and Saunders, performance is "a bodily practice that produces cultural meanings by 'translating' software scripts into an 'experienceable' reality" (ibid.). According to this, I understand the abilities of a robotic technology to interact, less as a fixed set of capacities, but rather as culturally-coded, bodily practice in a configuration of agents—that is, the more-than-human performance of robotic presents and futures.

With Suchman (2011), to study human-robot relations in practice encompasses reconstructing the practices of enactment through not only the "truncated labours of [the robot's] affiliated humans" (123), but also through "the figuration of subject object intra-actions in contemporary robotics" (121). Intra-action derives from Barad's (2007) account of *posthumanist performativity* and underlines the co-shaping relation between entities—instead of working with given, autonomous stable entities, namely the subject in opposition to the object. It depicts an onto-epistemological expansion of the analytical lens that takes into account that social practices shape material technologies, while they also become part of social practices and shape them themselves. Interaction at the human-machine interface does not follow a prescribed plan, but is the situated practice of intra-action which involves a set of sociomaterial conditions and a collective achievement of different sociomaterial instances that have to align in order to realise a certain goal (see Suchman 2007; 2011). This perspective depicts a shift from planning grasping as a movement which can be isolated, trained to and executed by a robot, and in turn always be legible in every setting and by every human as 'grasping', to understanding grasping as the result of such a collective effort that might rely on a host of different, not pre-planned movements and bodily experiences in order to be carried out between the in-practice differentiating entities of human and machine under specific sociomaterial circumstances.

Given these onto-epistemological insights into the nature of human-machine configurations of interaction as intra-action, I began to wonder how to open up possibilities for not only probing possible human-machine relations, but also enacting them *differently*. This centrally involved drawing on insights of FSTS in the process of imagining and realising a setting of HRI. My point of departure was the urge to scrutinise and move beyond hegemonic ideas of the universally useful robot that will become a co-worker or worker, taking over from 'us humans'. This entails to reflect the seeming inevitability of a robotic invasion through a more playful, experimental, and open engagement with one of the key players of such an invasion, namely the figure of the cobot. If I take the role of practices of enactment, such as the intra-active alignment of entities, experiences, and expectations seriously, what kind of setting and practices of relating and daring to enact human-machine-relations differently will become necessary?

What appeared intriguing as soon as I started contemplating about this, was to become curious about tweaking usefulness in HRI. Becoming curious describes an affective stance which I took on and which can be traced back to the very beginnings of AI research, for instance in Alan Turing's work, as Elizabeth Wilson (2010) has shown. More generally, affect plays a significant role in the relating of humans and machines. Sherry Turkle (1984) prominently analysed the human-machine interface as one in which machines are at the same time designed artefacts and evocative objects as they become interlocutors. The goal to create devices embodied with capacities to interact with 'us humans'—for instance through spoken or written language, but, depending on the artefact, also through gestures and morphological design decisions—is built on the idea of evoking affect in the affiliated humans, as also discussed in the preceding sections on robots as (co-)workers. However, beyond the endeavour to engineer affects as supposedly formalisable aspects for an increase in human-machine legibility, affects are also a core element of AI beyond such a formalisation. Wilson (2010) illustrates the ways in which the very foundations of AI are built on a curiosity about how thinking and feeling, as well as abstraction and embodiment, are deeply imbricated. In short, relations of intra-action between humans and machines are always affective. Precisely this affective quality articulated from the very initial moments of DRDK when I—amongst others of the MTIengAge team—was asked which scenarios we, members of the MTIengAge team, can imagine to be realising with the PANDA robot arms, I immediately said: "Knitting!" Then, this suggestion amused all of us, bursting out in laughter, including me, while at the same time, it started to fascinate us. We were immediately affected by robotic knitting and the idea to wrap our heads and hands, literally, around this playful challenge of implementing HRI through knitting.

I regard wool as a simultaneously disruptive and productive object in the setting of the robotic lab. Wool is a very stubborn material; wool can also be either very cosy and comforting, or itchy and unpleasant on the skin. As a knitter, I can tell that sometimes a specific yarn can break my patience when knitting, for instance, when the individual strings of the yarn unravel so that individual stitches start to dissolve and make it almost impossible to continue knitting. Of course, this depends on many sociomaterial circumstances, such as the quality and material of the skein and the needles, the knitter's experience and with it the embodied tacit knowledge around knitting, the comprehensibility of the used knitting pattern, as well as the lighting conditions. I give these details here in order to make tangible the scope, sociomateriality, experience-ability, and fragility of knitting. These multi-faceted dimensions of knitting as a practice are equally pivotal for establishing robotic knitting as those of human-robot interaction. Can knitting be transferred to a cobot? And: How can knitting become a test bed for probing the emerging cobot-human interface in practice, but also for re-crafting human-robot relations beyond this exemplary practice?

Clearly, bringing wool into a robotic lab at a technical university appears at first sight counter-intuitive. How and why a second look reveals a thread of long-standing overlaps between craft and technology will be the topic of the following chapter. Here, I want to dwell on the counter-intuitive momentum of the warm softness of wool and the practical character of knitting in the seemingly purely technological context, associated with cold metal and abstract programming. I regard this as the central interventionist momentum of robotic knitting: it disrupts the common understanding of proper objects in the robotic lab, while it also becomes an exemplary scenario of knitting thinking and feeling, as well as abstraction and embodiment together. In addition, while I, the FSTS scholar, had the idea of robotic knitting, it was at the same time important that the leader of the lab and roboticist, Raphael Deimel, could imagine this idea as not only potentially realisable, but also as a challenge for the lab's development of robot control. Thus, robotic knitting turned out to be an intrinsically integrative endeavour, crossing disciplinary boundaries.

Making the Queering of the Useful Robot a Matter of Care

Knitting with a robot amuses, as it appears to be an improper use of the robot and knitting needles, while this improperness invites me to pause for a moment and ponder about the very question of properness with regard to the implementation of HRC.

Bringing wool and knitting needles to a robotic lab as materials to engage with might cause irritation for several reasons. One quite obvious reason turned out to be the question of usefulness. Especially with regard to narratives of the cobot as becoming a worker, to work on realising the leisure activity of knitting with one's hands as a collaborative task with a cobot as partner does not contribute to narratives of useful cobots. Rather, the idea of realising a useless task with the coded-as-useful cobot stipulates an examination of the very notion of usefulness in the context of human-robot relations. These thoughts align with Sara Ahmed's (2019) *queer use*. Ahmed (ibid.,

2) delves into the question *What's the use?* as a "question of being" that points towards positionality and value. When Ahmed (ibid., 199) explores a potentiality in things that can be opened up through queer use, this entails to ask "how things can be used in ways other than for which they were intended or by those other than for who they were intended". Thus, with Ahmed (ibid., 7), "use is distributed between persons and things", while she also considers queer use to be *a start* for "making connections between histories that might otherwise be assumed to be apart" (ibid., 198). Bringing queer use into the context of the robotic lab, I regard robotic knitting as depicting such a queer use; a use which insists on the capacious nature of usefulness at the human-cobot-interface, and not only to queer what something is good for, but also for whom. Robotic knitting literally complicates use. With the situation in the lab in which I was asked to come up with a potential scenario of use, I was affected by the practicality of the robot arm, queering what a proper use supposedly should look like. In this sense, I consider the specific constellation of persons and things as foundational for stipulating the possibility to make not only a cobot, but rather equally the wool and knitting needles, in addition to the disciplinarily diverse situated subjects, the things and persons of queer use.

Moreover, queer use's emphasis on the potential and the distributedness of use can be brought into conversation with Maria Puig de la Bellacasa's (2017, 18) work on *Matters of Care*, for which "assembling neglected things" becomes central. More precisely, I suggest an account of queer use that involves queering as an activity of relating, and thus is constitutive of a method of caring for the neglected dimensions of an object's usefulness: its intrinsic potential to affect different uses and to transgress imaginable limits of what is considered to be the proper use of an object—in this case a set of objects, the knitting needles with wool and the robot arm. Hence, with Puig de la Bellacasa (ibid., 5), I understand care as "one way of looking at relations". Further, such looking at relations can involve both analysing the complex relations through which, for instance, human-robot interaction emerges, but also probing the very possibilities for relating differently. Care, then, is an affective stance of making the transgression of the limits of what is considered a useful use of the objects involved in robotic knitting, as well as what is considered proper knowledge and engineering practices in the involved disciplines, my matter of care.

While emerging robot technologies of a robotic future are ubiquitous, I also argue that they are at the same time elusive and intangible. It appears as if there exist a lot of ideas around 'our' robotic future, including automating the non-automatable, however, these narratives and images seem to mainly evoke pictures and suggestions on how 'our' robotic future might look. But how such interfaces will look like in detail and what it will actually be like to socially and physically interact with a robotic co-worker does not become tangible through most of these pictures. Thus, existing narratives mainly do not deliver insights into the nature of human-machine collaboration implemented through emerging technologies—especially its quality of a new, collaborative proximity between humans and robots. How proximate will these robots be and with which social and material consequences? As technology development has been heavily leaning on the graphical paradigm in computing (see Dourish & Bell 2011), for instance, embodied by the current models of PC, smart phone, and tablet, I wonder in which ways robots that clearly pertain to a different kind of machine than PC, phone, and tablet, are supposed to become part of 'our' technological everyday-environments. However, the inescapable and at the same time intangible quality of visions of 'our' robotic future makes it almost impossible to grasp the nature of action or even interaction attributed to emerging robotic technologies. Thus, from a performative perspective of a technofeminist intervention into the contemporary sociotechnical robotic imaginary, I am interested in exploring what I frame as the sociomaterialities of collaboration: that is, the practicalities, embodiments, and practices of relating.

String Figuring with Wool in the Lab: From Reflection to Diffraction

Bringing wool and knitting needles with me to the robotic lab depicts not only an intervention on the level of HRC. Rather, it is also an intervention on the level of knowledge and engineering practices, disciplinary expertise, and interdisciplinary processes. Reflecting taken-for-granted knowledge, it also raises the questions of who and what is included in designing and realising proper HRC, and who and what gets neglected. Finally, it deploys the conditions for an improper, queer use of cobots, knitting needles and yarn. This multi-faceted quality of robotic knitting is enabled by what I frame as engaging in situated practices of a co-engineering. Situating means more than adhering to the seemingly mere physical context of the robotic lab in which we (the persons with disciplinarily diverse backgrounds), the robots, the wool, and the needles met. Rather, it is about taking into account the sociomateriality of imagining, designing, and enacting HRC. It follows the impulse to situate knowledge claims, as it emerged from a feminist concern with the politics of location. Haraway (1991) popularly coined the term of *situated knowledges*. Suchman (2007) moves in a similar vein with her account of *situated action*. Situated knowledges and situated action are foundational to robotic knitting: while the first underlines the historically, socioculturally, spatio-temporally, and materially specific conditions of knowledge production, the second underlines the ways in which interaction between humans and machine results from a collective achievement that does not follow a plan, but the situational context. Situatedness here means, first, to take seriously the ways in which knowledge and artefact production cannot be cut off from their sociomaterial and political situation and context, second, a plea for a reframing of universal objectivity in terms of partial objectivity, and third, to become accountable for one's research in its reality-producing effects.

Moreover, in Haraway's (1997, 267) account of "situated practices or witnessing," she explains these practices of situating as analogous to witnessing, while she (ibid.) describes the latter as composed of forms of engagement with objects and knowledge production as "seeing; attesting; standing publicly accountable for, and psychically vulnerable to, one's visions and representations." My role in the lab became one of the *queer witness* (ibid.; see also Treusch 2015; 2017) who sees and attests the potential of human-cobot-inter- as *intra*-action for re-posing questions of automation, the organisation of work, robotic futures, and responsibility for technoscientific worldings.

A technofeminist, critical intervention then serves the purpose to both erode certainties and create further possibilities. With Haraway (1997, 95), "critical means evaluative, public, multiactor, multiagenda, oriented to equality and heterogeneous well-being." I understand this in terms of working towards visions of a robotic future which do not have to choose between a dismissive or an overly welcoming stance, but rather are guided by sociomaterial practice. Thus, the situating or queering witness is not to passively observe, but rather to become a part of the co-shaping of possibilities for making a difference in envisioning, representing, enacting, and becoming accountable for sociotechnical, robotic futures. It is precisely this role of the queering witness, to become accountable for the reality-producing effects as well as the partiality of observing, that is made possible by robotic knitting.

In Barad's (2007) continuation of Haraway's work on situated knowledges, she draws on Nils Bohr's quantum physics to further contemplate the relation between observer and observed. Barad (ibid., 196) basically asserts "that there is no unambiguous way to differentiate between the object and the agencies of observation". In this regard, Barad (ibid., 195) pushes further a shift from universal objectivity and representationalism to a "proto-performative account of scientific practices". Such a proto-performative approach foregrounds contingency, agency, and relationality, and necessarily involves tracing and engaging with the operations of power, namely, the dis- and enabling sociomaterial configurations of science and technologies. As Josef Barla (2019, 128) puts it vividly,

> "performative approaches circumvent the need for...a correspondence between world and words, matter and discourse by focusing on the question of how not only meanings but also particularly (re)configured bodies, identities, and hence realities, are enacted through particular generative practices."

In consequence, the matter of care in making a difference in 'our' robotic futures is not to find more adequate representations of what visions and realisations of collaboration between humans and robots really looks like, but to radically open up possibilities for new realities of human-cobot-relating—possibilities that might not have been explored before and that cannot be determined prior to probing them. The queering witness as the observer, however, can never be located outside of these generative practices of probing. In this regard, the boundaries of the persons and artefacts involved do not pre-exist their encounter, but rather materialise from this encounter. All involved entities are in a Harawayan (1991, 200) sense *generative nodes*—a term that accentuates the formative, co-shaping power of materialising locations in a net of actors.

Barad's account of the generative quality of sociomaterial practice is not only guided by Bohr's and Haraway's work, but also by Judith Butler's (1990) take on the performativity of gender and gender relations. In line with this, I work with the concept of performativity in the context of the robotic lab as a tool for acknowledging the powerful operations of social norms in the enactment of human-cobot relations—a tool which highlights that such a reproduction of norms holds the potential for an undoing of these very norms. Becoming a queering witness to the performative enactment of human-cobot relations is key in this.

It is again Barad's (2007) work and her expansion of the concept of performativity through a posthumanist lens, which has proven to be especially productive for researching human-robot relations in the making (Suchman 2011; Treusch 2015). This perspective allows me to understand every actor and artefact as a generative node, taking part in the crafting of robotic futures.

Hence, interaction is not regarded as the result of pre-planned affordances and constraints in the machine and (able-bodied, racial, gendered) norms in the human, but rather the result of co-constitutive relations emerging from a sociomaterial practice of relating between entities without pre-determined boundaries, which are immersed in and emerging from intra-active entanglements. These onto-epistemological insights into the performative, more-than-human nature of HRI make the processual, affective, and generative-subversive my matters of care—over the pre-planned, rational, and representational-normative—in understanding how humans and robots already do, but also could, relate in practice and in theory. As I will continue to show, this opens up the possibility to generate a more capacious understanding of HRC, including all actors and activities involved.

During HRI experiments prior to DRDK, I have witnessed many times that persons will hold their bodies in what I would call the outmost uncomfortable postures in order to increase the chances for mutual legibility between the robot, modelled as a child-like figure, and themselves, while also orchestrating their gestures, speech, and eye-contact, again, with the purpose to make the robot understand them. Not seldom, such situations were experienced and interpreted by the persons afterwards as exciting and pleasant, while I, as the observer, felt like my back was hurting just from watching and I assumed that having to go to all this effort would necessarily result in frustration. The positive evaluation of these persons can be read as emerging from an encounter between an evocative object and a person, in this specific case propelled and guided by the anthropomorphic, child-like robot figure. It limits the possibilities of relating on the one hand, and on the other disguises and neglects the labours invested in interaction in this setting. Clearly, the material, morphology, and behaviour of the machine play an important role, but also if and how the persons are given an instruction on what to do with the robot, the previous knowledge on the robot, personal and collective expectations of everyone involved, the setting of the room, the functionality of the robot on that day—all these sociomaterial circumstances condition not only how the HRI is enacted, but also how the self and the machine Other is experienced in intra-action. In this case, the machine Other becomes a capable interlocutor, embodying a set of capacities to interact. With Suchman (2007, 239), I understand this practice of enacting human-machine relations of sameness and difference as relying on practices of *enchantment* and *mystification* of the machine, installed by the human-like design of the machine, and as relying on the alignment of meanings and materials in this setting of intra-action, from which relations of sameness and difference materialise.

Returning to Brooks, the starting point in implementing my approach of a situated co-engineering through robotic knitting is the cross-disciplinary impulse to reflect on, in Brooks' words, where robotics is going and why. The wool then is a tool for tracing practices of enacting human-cobot relations, but also, and importantly, to open up possibilities to knot different strings together, forming generative nodes and opening up pathways of queer use which are to be explored. Here, the yarn's material and metaphorical meanings conflate. Especially, as the framework of co-situated engineering, as described here, also implies a shift from reflection to diffraction. Haraway (1997, 16) argues that "reflexivity has been much recommended as a critical practice, but my suspicion is that reflexivity, like reflection, displaces the same elsewhere." She (ibid.) continues to underscore that "what we need is to...diffract the rays of technoscience so that we get more promising interference patterns on the recording films of our lives and bodies.... Diffraction patterns record the history of interaction, interference, reinforcement, difference." Thus, diffraction allows stepping outside of the logics of finding the self in the Other, the mirroring relation of reflection, as well as the idea of a pre-existing ontology—existing prior to a reflexive practice. The second dimension becomes especially tangible through Barad's (2007, 72) work on a diffractive methodology as "a tool of analysis for attending to and responding to the effects of difference." As Barad (ibid.) further explains, this is about the "differences that our knowledge-making practices make and the effects they have on the world." Robotic knitting, hence, enmeshes different actors, discourses, and activities in order to establish a practice of situated co-engineering which does not rely on reflection, but on diffraction, as a method of inquiring into knowledge practices and not only practices of engineering HCI, but also of experiencing HCI. It is a diffractive practice of technofeminist intervention in its disruptive and generative momentum.

With Lykke (2010, 155-156), "diffraction and the cat's cradle game are two ways of describing an analytical process that is continuously innovative because of the ever-changing patterns of foregrounding and backgrounding, which aptly put the complexity of...objects at stake and thereby create new understandings of them." Thus, diffraction as a methodology is basically a generative practice of allowing new patterns of interference between different entities-as-nodes to emerge. Notably, in bringing the wool into the robotic lab, we were literally following strings of wool, dis- and re-entangling yarn with hands and grippers, and, in result, received (unasked for) patterns. This description not only depicts the everyday analogy for diffraction, but rather, in the context of robotic knitting, describes the very everyday practices of stipulating processes of a situated co-engineering through handling a yarn and knitting needles. The wool's stubbornness had an impact on how we worked, as well as on the pace of our progress in realising robotic knitting, as much as the stubbornness of the technical components (hardware and software), and finally, on the differing ways that individuals, as part of not-only-disciplinary collectives, enrolled in this process and experienced HCI. Every person involved in robotic knitting had to literally wrap their hands and head around the yarn, needles, robotic arm with user interface, and the practice of knitting, while these human engagements cannot be cut off from the more-than-human entities and sociomaterial circumstances involved.

Robotic knitting is my account of playing string figures with different sociotechnical, critical approaches, contemporary cultural imaginations of human-robot interaction, a host of materials like yarn, knitting needles, a robot arm, computers, and computer screens, as well as other persons. The idea behind such a string figuring in the robotic lab that necessarily involves engaging with a cobot technology is about re-crafting the story of robotic presents and futures. Notably, in its diffractive approach, re-crafting here is not about reflecting representations, but rather about diffracting entanglements. Finally, Daniela Rosner (2018, 56) reminds me that "the game of cat's cradle—composing string figures on multiple hands — ...is more than a form of illustration through storytelling; it is a material practice that makes way for new modes of being." Hence, situated co-engineering is about adhering to interference patterns as they emerge.

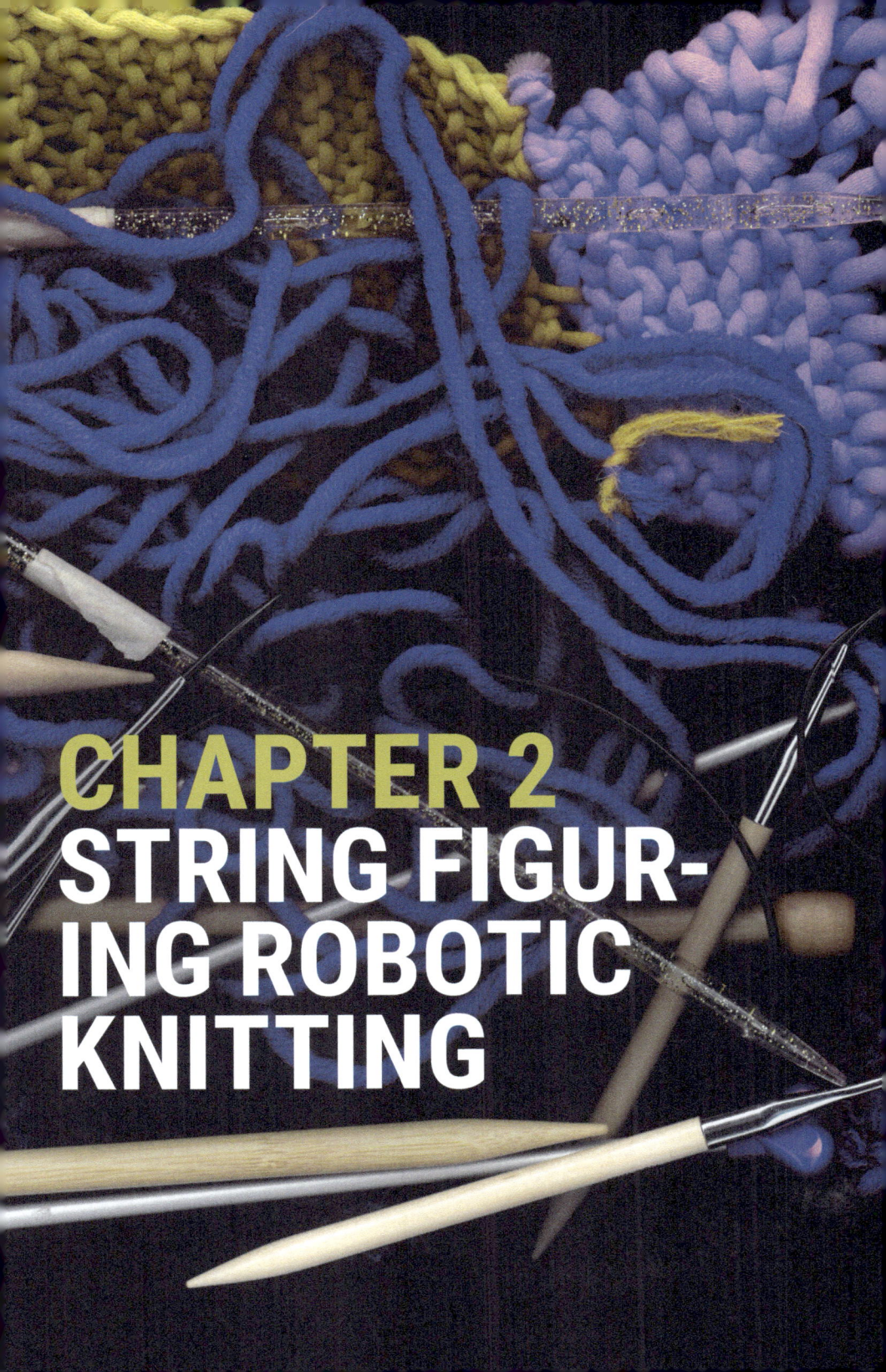

CHAPTER 2
STRING FIGUR-ING ROBOTIC KNITTING

Chapter 2: String Figuring Robotic Knitting

> The yarn is neither metaphorical nor literal, but quite simply material, a gathering of threads which twist and turn through the history of computing, technology, the sciences and arts. In and out of the punched holes of automated looms, up and down through the ages of spinning and weaving, back and forth through the fabrication of fabrics, shuttles and looms, cotton and silk, canvas and paper, brushes and pens, typewriters, carriages, telephone wires, synthetic fibers, electrical filaments, silicon strands, fiber-optic cables, pixeled screens, telecom lines, the World Wide Web, the Net, and matrices to come.
> *Sadie Plant, zeros+ones,* 12

Inspired by the twisted materiality of yarn, in this second chapter of the book, I will dis- and re-entangle what I identify as the two strings that are constitutive of my practice of bringing wool to the lab, playing with wool, knitting needles, and the robot arm PANDA, and realising collaborative knitting between humans and PANDA. The first string is formed by the historical entanglements of computational technologies and cultural techniques such as weaving and knitting, and the second string is formed through the collaborative intra-action between entities through the realisation of robotic knitting. Playing with both strings, I aim at allowing new patterns of interference between a cultural technique and cobots to take shape, as well as patterns of collaboration at the human-machine interface. Further, I am especially interested in recuperating historical legacies of such interferences as the basis for stipulating new stories of a technological future with a new generation of automata, working from the junction between handicraft and engineering as one of queer use.

Moreover, carefully assembling two strings, I engage in multiple string figurings of meaning- and matter-making between technologies and persons. In so doing, I not only make tangible existing relations between humans and machine as well as between digital and hand knitting practices, but rather also propel processes of relating differently. Hence, queering the usefulness of the cobot as co-worker, robotic knitting serves me as a tool for exploring the ways in which collaboration between human and machine is a sociocultural, material, embodied, and computational process, based on a host of meaning- and matter-making practices, and also always the result of a collective effort.

2.1 String 1: Knitting and the Digital—Diffracting Dichotomous Relations

The first string is about how the practice of robotic knitting not only centrally challenges common dichotomies between craft and high-tech, as well as virtual and tactile, soft and hard, and also female- and male-coded activities, but moreover challenges the very structure of ordering itself. The point of entry is the relation between knitting as a craft practice on the one hand, and digital practices of computing on the other.

My account of craft practices mainly encompasses spinning, weaving, and knitting. Looking at these practices in their contemporary meaning, they exist in both the private realm of reproduction and the public realm of production. In the first, they display a leisure activity or a hobby that one might have learned at school and pursues mostly at home. At the same time, all three craft practices can also be regarded as engines of industrialisation in the form of the invention of the spinning machine, the industrial weaving loom, and the (large scale) knitting machine. In contrast, the cobot used to be a rather large-scale, stationary technology that has been confined to the sphere of factory halls and by protective cages, while current waves of automation are increasingly pushing the engineering of rather mid-scale, light-weight, and flexible cobots as assistive machines supposedly entering the household (as figured, for instance, in the narrative of "the robots are coming").

As many social scientists have attested, we are living in times in which digital technologies, namely computer soft- and hardware, are ubiquitous as they have permeated every sphere of human everyday lives, constituting the conditions for a digital society (see Lupton 2015; Dourish & Bell 2011). In turn, social relations at work or in the private sphere, as well as forms of political participation, are becoming more and more polarised, either dis- or enabled through the use of digital networked devices. They basically stipulate a translation of analogue into digital data based on the binary coding of 0 and 1 and evolving through operations of formalisation and algorithmisation. In this regard, digital practices are not confined to a specific sphere of human existence, but are rather part of a comprehensive transformation of the sociomaterial grounds of human existence on both the level of cognition and embodiment.

Deborah Lupton's analyses of digital practices through the figures of the *Quantified Self* (2016) and, more generally, *Data Selves* (2019) both make tangible the ways in which people and digital devices are forming embodied "human-data assemblages" (Lupton 2019, 6). The notion of human-data assemblages underlines the ways in which "humans make and enact data" and the ways in which "data make and enact people" (ibid.). This co-creating relation then challenges ideas of the analogue and the digital as two separate spheres, limiting the relation between them to that of a one-way-translation in which everything becomes formalised and digital. Rather, with Lupton, I understand datafication as a much more lively process of entangling human worlds, devices, and spaces, constitutive of human-data assemblages from which (concepts of) bodies and devices with boundaries and agential capacities emerge. Returning to the quote at the beginning of this chapter, my interest in entangled human-data assemblages cannot be cut off from my interest in figuring out the role of the yarn for grappling with emerging, ubiquitous human-robot relations. What can be learned about technological histories, as well as futures by following and playing with yarn—not only in a metaphorical, but in a literal, material sense?

Shortly before the official start of the project *Do Robots Dream of Knitting?* (DRDK), I already made an appointment with Jan Martin, who was at that time the student assistant at the MTIengAge lab at TU Berlin. Jan is trained in electronic engineering and on that first meeting, we met to wildly associate knitting practices and robotics. This meeting was excellent for doing so, as Jan had not knitted before and I am not an electronic engineer. The yarn then took on the role of mediating between our disciplinarily diverse knowledge worlds, but also moved into the centre of our concerns with assembling ideas on how knitting with a robot could be realised. My very first idea, emerging during this meeting, was what I considered to be a simple transference of movements. It was based on my understanding of what this would imply, namely to conceptualise a sequence of movements which are definable through locations in space. We then just needed a model of the robot in space to which this data in turn could be transferred—at least this was how I imagined what robotic knitting would be about. At the time, I was only slightly familiar with the PANDA robot and motion planning methods. Nevertheless, this pathway of thinking and engaging with the yarn produced surprising knowledge and became one method of challenging taken-for-granted assumptions, including those I was working with.

During that day, I asked Jan to work with me on what I called the *datafication of a 'human knitting practice'*, namely my knitting practice. More precisely, we started with the aim of generating data on knitting movements by recording my practice of the garter stitch through motion capture. The robotic lab was endowed with eight infrared cameras of a motion-capture system. In short, motion capture records the movement of markers (mostly small polystyrene balls covered with a retroreflective material) in a defined space that is recorded from different angles through several calibrated infrared cameras, which together produce a 3D image of the markers moving in space on a screen. In classical human movement analysis, the three markers have to be placed at a selected joint of a human body, representing the x-, y- and z-axes in space. Using a MoCap System body model, these markers then allow one to track the orientation and position of each joint in motion. In the case of knitting, I decided that the goal was not to understand how my arms and hands with fingers are moving—as the robot arm is endowed with a gripper and not with a human-like hand—but rather to track the movement of the knitting needles. If we could locate the exact position and orientation of each needle during the different movements of, for instance, the garter stitch (knit every row), wouldn't it be possible to emulate this behaviour of the needles when the needles are put in the gripper of a robot instead of a human hand? Moreover, my assumption was that the garter stitch, in its repetitive qualities, should produce a pattern of movements in space that are marked by a high degree of uniformity and would therefore be easily reproducible by the precise cobot arms of the newest generation.

Against this backdrop, Jan, Bülent Erik (the head of the technical staff), and I started to think about where the markers should go on the needles. The next step was to figure out how to attach the markers onto the needles. We attached three markers on each needle, representing the needle's location on the x-, y-, and z-axes. We also used the standard method of attaching markers, which is the use of Velcro tape segments to which plastic threads are mounted so that the accompanying markers must only be screwed onto the thread. On this first, more playful occasion, I only brought one pair of needles with me: bamboo knitting needles size 8.

However, the surface of the bamboo needles turned out to be too slick for the underside material of the Velcro tape, so that the tape with the marker could not be placed in a fixed position. Luckily, Bülent came up with the idea to use some crepe tape underneath the Velcro tape to make the needles larger in their diameter and to increase their skin friction. *Figure 1* shows the needles with three markers.

Figure 1: The needle with the markers

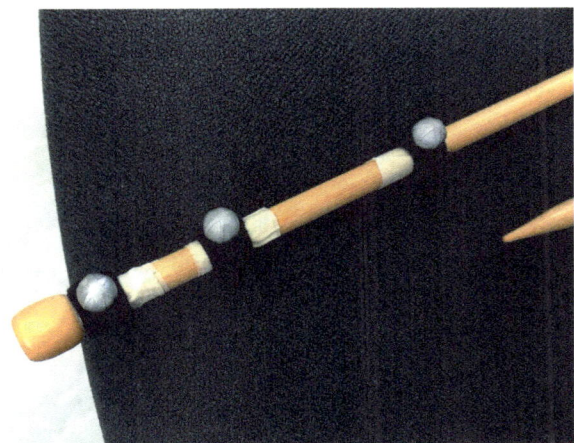

After the needles were prepared, I started to knit, while the MoCap System tracked the movements of the needles. The generated data can be displayed in various ways. I was fascinated looking at the numerical version (*Figure 2*), illustrating the exact location of each marker in the coordinate system during each second. My knitting practice never looked like this before.

Without even knowing how to read these numbers properly, their variations nevertheless already challenged my assumption about the uniformity of

Figure 2: Screenshot of MoCap Data

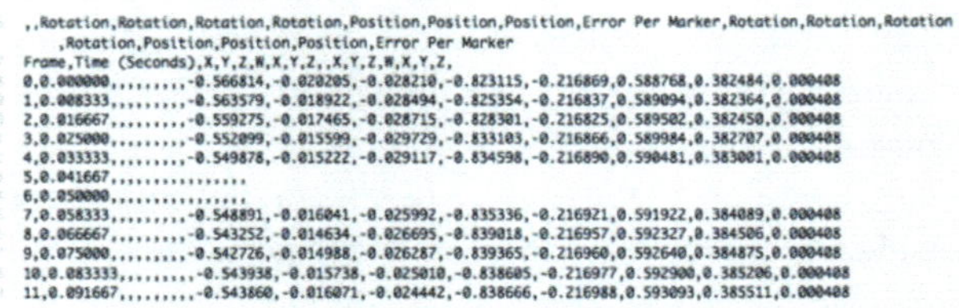

```
,,Rotation,Rotation,Rotation,Rotation,Position,Position,Position,Error Per Marker,Rotation,Rotation,Rotation
,Rotation,Position,Position,Position,Error Per Marker
Frame,Time (Seconds),X,Y,Z,W,X,Y,Z,,X,Y,Z,W,X,Y,Z,
0,0.000000,,,,,,,,,,-0.566814,-0.020205,-0.028210,-0.823115,-0.216869,0.588768,0.382484,0.000408
1,0.008333,,,,,,,,,,-0.563579,-0.018922,-0.028494,-0.825354,-0.216837,0.589094,0.382364,0.000408
2,0.016667,,,,,,,,,,-0.559275,-0.017465,-0.028715,-0.828301,-0.216825,0.589502,0.382450,0.000408
3,0.025000,,,,,,,,,,-0.552099,-0.015599,-0.029729,-0.833103,-0.216866,0.589984,0.382707,0.000408
4,0.033333,,,,,,,,,,-0.549878,-0.015222,-0.029117,-0.834598,-0.216890,0.590481,0.383001,0.000408
5,0.041667,,,,,,,,,,
6,0.050000,,,,,,,,,,
7,0.058333,,,,,,,,,,-0.548891,-0.016041,-0.025992,-0.835336,-0.216921,0.591922,0.384089,0.000408
8,0.066667,,,,,,,,,,-0.543252,-0.014634,-0.026695,-0.839018,-0.216957,0.592327,0.384506,0.000408
9,0.075000,,,,,,,,,,-0.542726,-0.014988,-0.026287,-0.839365,-0.216960,0.592640,0.384875,0.000408
10,0.083333,,,,,,,,,,-0.543938,-0.015738,-0.025010,-0.838605,-0.216977,0.592900,0.385206,0.000408
11,0.091667,,,,,,,,,,-0.543860,-0.016071,-0.024442,-0.838666,-0.216988,0.593093,0.385511,0.000408
```

the knitting movement. The rows show similar numbers, but they are never identical—they all vary from each other. If the movement of, for instance, inserting the right needle into a stitch on the left needle in order to open up a new stitch, repeats itself approximately every five seconds, then this movement should have the same numbers in every fifth row. Even more revealing is watching the MoCap animation of this data, as the following screenshot (*Figure 3*) shows:

Figure 3: Screenshot of recorded MoCap tracking

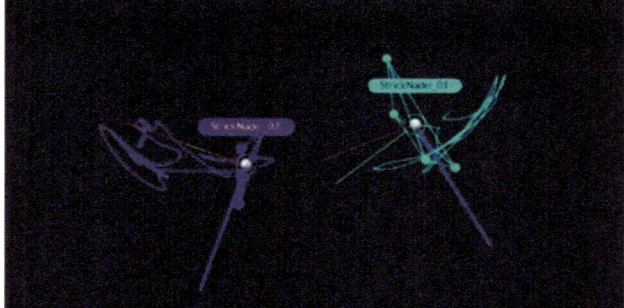

Needle one and needle two (*Stricknadel*) perform movements that do not reveal any kind of obvious rhythm and cannot be identified without an explanation. They show no uniformity at all. At a first glance, the MoCap captured movements that could be loosely associated with hands conducting an orchestra. This visualisation made me pay attention to the myriad of micro-movements that I constantly make in order to control the behaviour of the yarn, the stitches on the left needle, the new stitch on the right needle and the knitted piece. These movements of compensation and correction prevent the individual stitches from slipping from the needles, but also secure an even length of the stitches and therefore uniformity of the knitted piece. Clearly, the practice of knitting with two knitting needles is very different to the automated form of knitting performed by knitting machines.

These initial insights into the nature of 'human knitting' already displayed the different layers of obstacles that we would probably have to encounter in the following course of the project and in its attempt to transfer my knitting movement onto a robot arm. In the case of knitting with two needles, it is essential to note that the behaviour of the yarn turned out to be unpredictable. This, in consequence, also made the behaviour of the yarn and the movements of the needles when hand knitting non-formalisable. In contrast, my vision of automating 'human knitting' relied on producing a data set of replicable movements that in turn can be transferred onto the robot. Nevertheless, I was still—if not even more—intrigued by the idea of having a robot arm knit collaboratively with me.

Motion capturing my knitting practice not only illustrates the unexpected obstacles, but also makes tangible the sociomaterial, embodied circumstances of the digital practice of MoCap. In order to track the knitting needles, we not only had to endow the needles with markers, but I also had to position myself in space in a way that I was visible to every infrared camera and I had to fixate my body in a way so that the needles were performing only the knitting movement. Basically, I was sitting on a chair, bent forward with my arms resting on my upper legs while knitting. This was a very uncomfortable position and definitely not one I would have chosen for knitting if not for the MoCap tracking. I had to practice this form of uncomfortable knitting to prepare myself for the MoCap tracking. Despite my best effort to make 'human hand knitting' datafiable, we did not receive the expected results: A reduction of complexity of the movements through formalisation as a key to motion planning. Rather, this attempt at datafication challenged the very idea of knitting as a repetitive and therefore uniform practice, making tangible the material resistances and stubbornness implicit to forming new stitches.

Moreover, though not producing the expected result, I nevertheless contemplate this first attempt at translating my hand knitting into a plannable motion through MoCap tracking as producing an interference pattern which opened up the possibility to dis- and re-entangle yarn, knitting, and digital practice differently. Further, this result raised questions, such as: How is it possible to mediate between 'human knitting' and the promise of automation carried by the core strength of the latest-generation flexible robot arm? That is, how can we reconcile the robot's absolute and reliable repetition of the same movements with outmost precision if knitting is a practice characterised by non-uniform movements? And, what kinds of human-robot collaboration would emerge from this? Clearly, datafication in this case opened up the need to delve deeper into the qualities of mastering knitting as a technique, as well as the legacies of knitting, challenging my under-complex and very reduced initial image of hand knitting and how it relates or can be related to digital practice.

My first encounter with knitting was during my PhD studies. During that time, I would often get frustrated with academic work, the endless hours spent in front of my computer, mostly at home as I did not have a proper office space at my disposal at the time, while feeling at the end of the day that the outcome of the day's work was not really tangible to me. The size of the document files, but also the number of sometimes handwritten notes might have been growing, but I nevertheless rarely felt satisfied with what I had accomplished. In addition, I was also frustrated by the structure of such a day, as I experienced writing my dissertation at home as a process which could take hours of reading, revisiting notes and yet not producing a countable output, namely written pages, while this process could easily stretch into the late evening, transgressing the boundaries of work into my supposed leisure time. Burdened by these circumstances of writing, I started to search for an activity that could balance the writing process and result in a more tangible outcome. By chance, knitting became this activity for me.

At first, I started to learn how to knit socks through a knitting book, but quickly needed to complement the written guide with online video tutorials, as I needed moving 3D explanations of where needles and yarn have to be in order to form new stitches and to understand how I can get them there. Mastering the learning and acquiring the tactile skills needed, the regular evening practice of knitting socks soon meant my involvement in a repetitive, rewarding activity, which accompanied the process of my dissertation writing nicely.

Knitting as a handicraft leisure activity depicts a practice of structuring, involving ordering as well as creativity and meditation. Knitting is the goal-oriented activity and creative practice of producing new stitches and thereby also a knitted artefact, while the repetitive nature of forming new loops requires focus and is at the same time meditative (Wiescholek 2019, 73). It is also a satisfying practice as my experience confirms: The finished socks as the knitted outcome are "a physical manifestation of a knitter's effort, skill, and productive use of time" (Rosner & Ryokai 2008, 2). Thus, knitting as a practice is more than mastering a technique—it also involves the stipulation of different senses, affective states like comfort and satisfaction, and mental activity. The resulting piece will show traces of this embodied, multi-modal, and mental practice. As Daniela Rosner and Kimiko Ryokai (2008, 1) underline further: "A handcrafted artifact can physically embody the skill and time involved in its production. For example, the subtle unevenness of stitches in a hand-knit textile may be an indication of the rhythm and tension of the knitter at that particular point in time those stitches were created." A change of rhythm and tension in knitting socks, for instance, might cause me to produce a pair of socks that differ in width and size, though made with the same number of stitches and rows. Every knitted artefact embodies the history of its production, including the persons involved, the locations of knitting, the skills of the knitter, as well as the varying executions of the planned stitches. The knitted artefact is a manifestation of this story.

Furthermore, an essential part of learning how to knit was learning to differentiate between knitting styles, English and Continental (basically, the yarn is held either in the left or in the right hand), and deciding which one I would like to learn. Next step was to learn not only the movements of my two hands with the two or, when knitting socks, four knitting needles, but also to learn the formal language of knitting patterns. Patterns are written in a code that mainly differentiates between the two stitches knit (k) and purl (p). The basic and important stocking stitch alternates between a knit and a purl row, but as socks are knitted in rows, every round is a knit round. Another very basic pattern which I also learned very early on is the moss stitch, which can be used to add a different texture to the socks, and which requires four alternating rows. The knitting pattern for the moss stitch over 4 rows looks like this:[1]

Figure 4: Knitting language

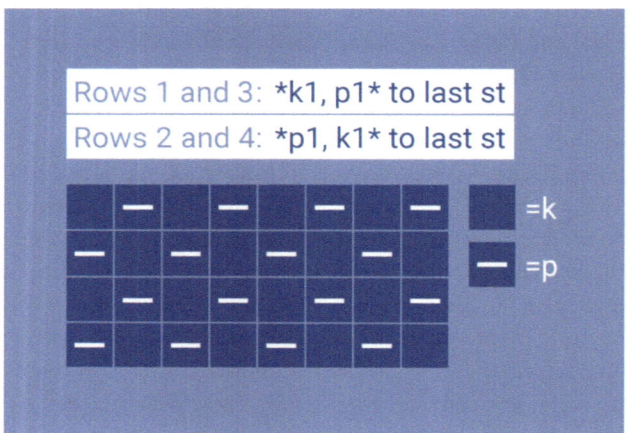

What immediately stands out is the binary structure of the knitting language. This binary logic is at the core of any kind of (textile) web. Tracing the handling of yarn even further, it is interesting to dwell a bit further on the binary logic of knitting. Hence, in its structuring quality, the binary code of textile is also a pivotal point for a classical, utopian cyberfeminist position. In short, this position emerged at a specific historical junction during the 1990s and "their themes of 'grrrl power' and 'wired worlds'" (Wajcman 2004, 63). It is relevant to situate cyberfeminism within this junction and—in line with Wajcman (ibid.)—as "a new relationship between feminism and technology" that foregrounds empowerment through female agency and subjectivity. For one of the most popular figures of cyberfeminism, Sadie Plant, this new relationship is centrally connected to the binary code in its ambivalent meaning: It represents both the Western patriarchal symbolic order and the potentiality for female empowerment and emancipation.

---

1   *Figure 4* shows the written pattern on the left, and on the right a translation of the pattern code into a chart. The asterisks indicate that the instructions between them have to be repeated—in this case, until the last stitch (last st) of the row.

Plant (1998, 34-35) emphasises that "the zeros and ones of machine code seem to offer themselves as perfect symbols of the orders of Western reality...the difference between...form and matter, mind and body, ...inside and out, active and passive, ...yes and no, ...male and female." Wajcman (2004, 35) further explains: "It takes two to make a binary, but...1 and 0 make another 1. Male and female add up to man." Haraway's figure of *Universal Man* literally represents this binary logic. However, Plant also differentiates between such a binary logic and the operations enabled by digital technologies, ranging from the (historical) example of punch cards to the modern, networked computer and how women have been and are still involved, as well as the ways in which binary identities become flexible. Thus, she understands these operations primarily as holding the potential to transgress if not subvert a binary logic in its powerful operations of differentiation and classifications. Plant explains this by drawing on technology as the compound of *techne* (art, skill) and *logos*. Further, following Plant (1998, 50), the latter implements the logic of the binary system, differentiating between the one and the other, which is not one (the zero), while the first indicates that technology is "also a matter of the skills, digits, speeds, and rhythms of techno." As she continues: "The techno and the digital are never perceived to run free of the coordinating eyes and hands of logic and its binary codes. But logic is nothing without their virtual plane" (ibid.). After all, the organising unit of the binary digit (bit) is the byte, eight bits. Plant (ibid.) understands this pairing of eight as the basis of digital technologies "full of intensive potential." With this, she re-codes not only the relation between women and technology as inherently intimate, but also between zeros and ones, as "zeros now have a place, and they displace the phallic order of ones" (Wajcman 2004, 64).

Technofeminists, cyberfeminists, and FSTS scholars, more generally, have written extensively on Plant's vision of a feminist cyberculture and cyberfuture, especially her blind spots regarding the exploitive and oppressive dimensions of the cyberverse and her essentialising and determinist tendencies (see Wajcman 2004). However, and as Cornelia Sollfrank (2018, 12), another central technofeminist voice, underlines, such a critique of early cyberfeminisms is in danger of failing to acknowledge the *Wirkmächtigkeit* of the political fantasies evoked by the very notion of cyberfeminism. Thus, reading Plant's take on zeros and ones evokes the possibility to speculate about a transgression of binary logics through a different re-enactment of the ordering structure of the logos in technology. Zero and one then always embody the order (*logos*) and the virtual (*techne*), bringing both into a relation of tension and interference rather than opposition and dichotomy. Plant's work depicts a unique take on yarn as not only metaphorical, but a material component along which the history, present, and future of technological innovation takes shape. Thus, playing with yarn in a robotic lab not only uses the challenge of handling a yarn to disrupt the narrative of "the robots are coming" and generates the possibilities for new stories. Rather, it also moves in such a vein of working through the tension between order and subversion, evoking transgressive potentials, re-joining fields of meaning and of practice, as well as re-connecting histories, presents, and futures. I understand this conjunction of meaning and material in the yarn in terms of the game of string figuring, which diffracts and is performative at the same time: It allows patterns of interference between yarn, knitting (technique), and technology (*techne* and *logos*) to emerge, while iterations of producing structure also always enable new patterns to emerge and to be enacted.

In the case of knitting, the binary basis of the knitting syntax allows complex patterns for various texture effects, such as lace, the intended production of holes. For instance, the diamond knit lace produces holes that form a diamond in the knitted piece. In order to knit such a diamond structure, one has to constantly increase and decrease the number of stitches through the following techniques: knitting or purling two stitches together (k/p2tog = knit/purl 2 together), slipping and stitching over (skp = slip1, knit1, pass slipped stitch over), yarn over (yo), and knitting one or more of these stitches through the back loop (tbl). The diamond lace pattern in a chart looks like this:

Figure 5: The diamond lace chart

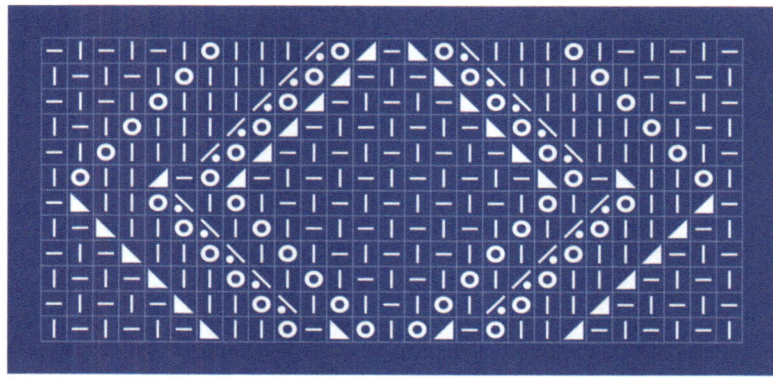

This chart helps the knitter to orient themselves in the pattern as it visualises the written code (row 1: ... p1,k1,skp, k1, k1, yo, p, skp, yo,k1, yo, k2tog, ...). It does so by dividing the diamond structure into rows and stitches and allocating the x- and y- axes to them, so the different lace techniques are mapped onto a position on these axes. Moreover, displaying the knitting code in a chart, the chart shows similarities with a punch card. Historically, the punch card originated as a mechanism for improving the work at the weaving loom at the dawn of the 19th century. This weaving mechanism was named after its alleged inventor, Joseph Marie Jacquard.[2] Even though the invention of mechanisms like that of the Jacquard loom were experienced as a threat to the weavers as it "withdrew control of the weaving process from human workers and transferred it to the hardware of the machine," which was further experienced as a process in which "a piece of [the workers'] bodies literally being transferred to the machine" (Plant 1998, 15), this mechanism had a different intent: It was targeted at increasing the artistic quality of weaving. As Birgit Schneider (2007, 295) explains:[3]

> "The punch cards promised the potential to improve the craftsmanship of the silk weavers in order to increase the quality of the fabric, produce more beautiful designs, highly skilled colour nuances and more manifold light effects while simultaneously enormously save on time and in costs."

Thus, what can be noted here is that the modularisation of weaving through the Jacquard mechanism is a very vivid example for a transference of a handicraft practice from human workers to machines. Intriguing to me is that this *automation* was feared to cause humans to not only give up control over their structure-forming and creative craft practice, but to even give up a part of their bodies, while the *mechanism* itself was—when developed—thought of as more as an extension of the weavers' abilities to craft and re-weave.

As Schneider (2007, 294) reminds us, Jacquard's punch card mechanism invented the loom as a "meta-machine" by "dividing the control mechanism from the information." This, moreover, became a point of departure for Charles Babbage and Ada Lovelace to build their model of the *Analytical Engine* in 1838. Celebrated as heralding the modern computer, the Analytical Engine,

---

2   For a detailed account of the development, implementation and improvement of the Jacquard mechanism, see Birgit Schneider *Textiles Prozessieren* (2007).
3   Selected quotations from Schneider's (2007) book *Textiles Prozessieren*, published in German, have been translated to English by Pat Treusch.

as Lovelace (Lovelace cited in Plant 1998, 18) writes, uses "the introduction of the principle which Jacquard devised for regulating, by means for punched cards, ...rendered it possible...to make this engine the executive right-hand of abstract algebra." What intrigues me here is that the Analytical Engine is thought of as becoming a help in algebraic operations, figuring as the *executive right-hand* to 'us' humans. Thus, Lovelace imagined the Analytical Engine to work as a complementing device in analogy to the Jacquard weaving loom. Both mechanisms are based on the punch card as a "storage medium and control module," while the Analytical Engine processes given information by computing an output, the Jacquard loom processes information by realising a "woven image" (Schneider 2007, 298).

Returning to the case of hand knitting, the chart displays information that functions as a helping guide for the knitters' hands. Even though it visualises the different knitting techniques—like p1, k1, skp, k2tog—to navigate through such a complex pattern not only requires to be able to master each technique, to read the code, and to understand the chart of the pattern, but also to attain a sense for the structure of the knitted piece—in this case, the diamond lace (see *Figure 5*). This encompasses developing tactile knowledge of each technique so that the knitter knows how the movement to knit the skp stitch, for instance, feels as well as the structural haptics of the texture produced by this technique in the knitted piece. The visual information on stitches is only one dimension of orienting oneself when knitting such a complex pattern.

The example of the diamond pattern makes tangible the ways in which following such a pattern is—through its repetitiveness and complexity—at the same time a focused and meditative activity of producing a uniform structure with repeating elements in which the unmaking of certain stitches and the making of others provokes the creative moment of the diamond lace materialising. The use of the different stitch techniques is goal-oriented towards producing structure and creating the diamond lace as a set of specific stitch techniques. Sybille Wiescholek (2019, 52) points out that the structure-creating properties of patterns for textiles encompass two dimensions: "The textile creates structure in the human and the textile activity as handicraft is at the same time a mental activity."[4] Here, she brings together the dimension of using a pattern to create a material structure from a thread and the dimension of human patterns of thought and action that reveal themselves in the

---

4   Selected quotations from Sybille Wiescholek's (2019) book *Textile Bildung im digitalen Zeitalter*, published in German, have been translated to English by Pat Treusch.

textile. Knitting can be understood as both a craft and intellectual activity, while patterns of making and patterns of thinking both are based on their structuring operations. In addition, Lydia Maria Arantes (2017) in her cultural anthropology of knitting as *Verstrickungen* (enmeshment)[5] between *interior spaces*, *women's spaces*, and *economic spaces* (321), develops an account of hand knitting as a technical activity that combines corporal, material, mental and sensual practices (ibid., 86).

Drawing on the complex nature of knitting as a technique, the process of forming a textile structure then can be tweaked through a diffractive lens: The iterative dimension of knitting as well as the way in which norm and subversion interfere through knitting, both depict a way of knowing and enacting structure. These complex and compound meaning- and matter-making dimensions of textile practice also show in the use of string figuring as an everyday analogy for diffraction and my account of knitting in the context of robotics as a practice for challenging existing orderings of human-machine relations to, not least, trace subversive impulses. In short, robotic knitting can be regarded as initiated by a curiosity about the relationship between sociotechnical orderings, regulating how humans and machines can relate, and knitting with a yarn can be seen as an embodied, materially and intellectually structuring, as well as affective practice of ordering and at the same time subverting order. Could a look at the history of knitting reveal its relevance as a cultural technology of structuring? And if so, what can I learn from this about dichotomous relations between craft and digital practice, but also between male and female, and human and machine?

According to Ebba D. Drolshagen's (2017, 25) account[6] of a Western and mostly European cultural history of knitting, there exists no proven knowledge of the beginnings of knitting. The oldest found and dated knitting artefacts, are a pair of "natural white knee highs, stocking stitched and apparently round knitted, with perfect heel, perfect toe, a fit following the calf and a precisely worked, intricate knit-in pattern in indigo blue" (ibid., 28). They are dated back to the 11th to 13th century AD. Due to their delicate makeup, these artefacts clearly do not witness the beginnings of knitting, but rather testify

---

5 Selected quotations from Lydia Maria Arantes' (2017) book *Verstrickungen*, published in German, have been translated to English by Pat Treusch.
6 Selected quotations from Ebba D. Drolshagen's (2017) book *Zwei rechts, zwei links - Geschichten vom Stricken*, published in German, have been translated to English by Pat Treusch.

that "they are the result of a long learning process" (ibid.). What is known is that, from the 13th century on, the first hand-knitter guilds were founded by professional knitters throughout the area of Europe, while, as Drolshagen (ibid., 33) emphasises, "at least throughout its heyday, ...only men were part of the guilds." The boundaries of a proper knitting practice were regulated through its professional organisation in guilds, while the members were almost exclusively male and the practice therefore male-coded.

Jumping to the 19th century, knitting machines became increasingly relevant, while to operate knitting machines advanced into a "purely male profession" and hand knitting as a profession lost its meaning and became "an exclusively female occupation" (ibid.). Tracing such a gendered coding shows how Western symbolic order and cultural technologies are intertwined. As Jack Z. Bratich and Heidi M. Brush (2011, 235) underline, it is the simultaneity of "an economic reorganization of the bodies of male workers" and "the dispersion, deauthorization, and expropriation of women's skills and knowledges along with the destruction of many women's bodies" that is foundational for the interrelated industrialisation of handicraft, the rise of capitalism, and concomitant gender hierarchies. Thus, in what follows, I will present selected historical stations of hand knitting with an emphasis on the structuring nature of knitting, exploring the latter as a momentum of change inherent to the practice of knitting in its quality to reproduce order while carrying the possibility to subvert order.

Though not a proper profession any longer, hand knitting nevertheless is work, especially amongst (sheep) farmers and workers and until the late 20th century, also including every step of yarn fabrication. Here, knitting becomes something to ensure the survival of families. Knitting as existential labour was done mostly by females (women and girls), but also more generally among children as young as three years old. Drolshagen reports that working-class women and women on farms were constantly wearing a small bag with them, even when already working, for instance, as peat cutters, so they could additionally knit throughout the day. She (2017, 71) explains that "a walking knitter...needed 15 kilometres to finish a sock." At the same time, the knitters were dependent on retailers and were mostly exploited. Thus, Drolshagen's (ibid. 77) comparison of hand knitting as labour to *compulsory labour* seems adequate. In German and Austrian regions, compulsory labour was also called *robath, roboth* or *robot* from the Slavic *robota*. In this sense, the women and children had to turn themselves into *knitting robaths* to secure their existence. Hence, the practice of knitting in the reproductive realm at that historical period was far from a pleasant leisure activity.

In addition, in bourgeois circles of the 19th century, knitting became a female virtue. Women were supposed to knit during the day, which should avoid idleness and the danger of falling into sinful behaviour like eating (excessively) or feeling sexual desire (ibid., 107). Knitting served the purpose of disciplining female bodies in line with the gender and class orders and female role patterns of the social elite of that period. Women had to, at the same time, represent the wealth of the family and always be active, but not working (ibid., 109). The constantly knitting, embroidering, or crocheting hands of women and girls can be regarded as emblematic for embodying this ideal. However, and notably, it was also important to not produce anything of use, as Drolshagen (ibid., 110) points out: "Only small, fine, decorative things came into question, which could be held delicately in the hands and did not ruin the femininity of the silhouette."[7] The active hands of the modest females, however, also became a bone of contention for women to start to emancipate from this idealised role. On the one hand, there exists the historical example of the *tricoteuses*, the female knitters with varying class backgrounds, who during the French Revolution knitted in public (mostly during executions) as a mode of female political participation. On the other hand, women also began to revolt against this female role pattern in its condemnation to sit still and knit, crochet, or embroider apparently useless things (ibid., 118-19). In short, the knitting practice is a substantial dimension of the powerful operations of establishing and reproducing normative social orderings and of an emancipation of women of exactly these normative intersectional gendered, classist role patterns.[8]

During both World War I and World War II, knitting became a patriotic act and women and children were called upon to produce woollen socks, hats, and sweaters for the soldiers, turning hand knitting into a "labour of love" and a "service for the fatherland" (ibid., 122). Parallel to this, knitted clothing slowly became fashionable from the 1920s on—from Coco Chanel's famous jersey fashion to hand knitted sweaters (ibid., 148-150). The advancement of

---

7   As Drolshagen (ibid., 144) also notes: "That [idealised role distribution] was often only a façade." Women also in bourgeois households often had to make extra money with their secretly made handcrafted pieces that therefore necessarily needed to be useful, thus, marketable.
8   For an analysis of historical legacies of global female textile networks as a *catalyst of change*, see for example Sinclair 2015.

knitwear into fashionable clothing, which women could produce for themselves, can be mainly regarded as foundational for the rise of hand knitting as a female-coded hobby throughout the 20th century with its continuation into the 21st century. Clearly, this hobby could turn out very useful during times of financial crisis and also bridge gaps in supply in the 20th century. However, the female hand knitting practice was nevertheless mostly disguised as a hobby, the necessity to knit instead of being able to buy clothes was neglected and knitting became a substantial part of the unpaid and unacknowledged reproduction work provided by women (ibid., 151). At the same time, hand knitting as female-coded practice co-existed as a source of income for women and families from around 1930 until the 1980s in the form of home-based work.

Drolshagen nevertheless underlines the difference between hand knitting as a leisure activity, despite being mostly unpaid labour, and as paid labour, while the latter forced the knitters (women and men) to work efficiently and very precisely. As she (2017, 159) writes: "The fixed gaze towards the market ultimately degraded knitters to breathing hand knitting machines." This description of the working conditions of hand knitters makes tangible the ways in which knitting is an activity that can be automated on a massive scale as it historically has been, but also that there are limits to having machines produce knitwear. In consequence, it seems like hand knitters have had to turn themselves into machine-like producers throughout different historical periods. This process, furthermore, is regarded as reducing hand knitting to the production of knitwear and thus in danger of erasing the creative, but also technically challenging, dimensions. It appears as if the one (machine automation) would have to exclude the other (human creativity) by necessity. Could hand knitting and machine practice be re-joined differently? And if so, through which sociomaterial circumstances and practices?

Nowadays, hand knitting has advanced from a hobby pursued privately at home into a quite fashionable leisure activity, which is widely represented publicly, for instance through knitting circles, craft fairs, and growing numbers of books and magazines, and as organised via the internet. The internet provides a space for connecting with other knitters, sharing knowledge, for instance, on how to knit (which I profited from, too), and potentially for designing and selling knitting patterns, but also handmade knitwear. The most well-known pattern platform is *Ravelry* (www.ravelry.com) and there exists an uncounted amount of knitting and crocheting blogs. One of these bloggers, Stephanie Pearl-McPhee, has coined the term *interknit* as a term that signifies the relevance of "the internet for the global knitting community" (Pearl-

McPhee, cited in Drolshagen 2017, 174). The interknit can be recognised as foundational for different waves of contemporary knitting activism, so called craftism or craftivism—an activism, which is built on craft as a motor for social change.

For the first decade of the 21st century, "an explosion on the popularity of knitting" (Springgay 2010, 111) can be attested. The scenes in which this explosion takes place are multifaceted: Beth Ann Pentney (2008, 1), for instance, locates "an upsurge in Western popular culture," a phenomenon that has been analysed as leading to *celebrity knitting* (Perkins 2004), but is also especially articulated in the form of reclaiming knitting as a feminist, craftist practice in all its paradoxes between empowerment and consumerism (Pentney 2008; see also: Kelly 2013; von Bose 2018). One phenomenon amongst this reclaiming is that of college-educated, white Western women who are quitting their jobs in order to knit or craft and sell their products online. Here, reclaiming knitting seems to be tied to the emergence of a *new domesticity* which is thought of as serving this particular group of women—in line with Käthe von Bose (2018, 198)[9] —to strive for "personal fulfilment through a more natural, more conscious lifestyle with handicraft." Furthermore, von Bose (ibid.) identifies this phenomenon as paradigmatic for the rise of an *aesthetic capitalism* characterised by the neglect of "the complex interconnections and reciprocal conditionalities between creativity and market logic." However, and as von Bose (ibid., 199-200) further underlines, reducing this form of knitting activism to an articulation of aesthetic capitalism fails to take into account "the ambivalent elements of concrete practices of producing DIY clothing...for instance, that traditional handicrafts which are based on haptic, materiality and embodiment, are merging with technologies of digitalisation and virtuality." Decades after the emergence of cyberfeminism, the yarn in its metaphorical and material meaning for the re-formation of social order as well as the shaping of technology seems to regain momentum, leading to a re-joining of handicraft and digital practice in a transgressive manner.

Precisely this relationship between handicraft and digital practice is also at the core of Daniela Rosner and Kimiko Ryokai's (2008) and Rosner's (2018) work on the design study on *Spyn*. Spyn in short, is a digital device designed

---

9   Selected quotations from Käthe von Bose's (2018) article ‚*Mit Liebe handgemacht'. Nachhaltige Do-it-yourself-Mode als körperlich-affektive Geschlechterpraxis*, published in German, have been translated to English by Pat Treusch.

to expand a knitters' possibilities to document and share their knitting practice digitally while using the device to store memories and stories digitally at selected locations in the artefact. Rosner (2018) captures the beginnings of the interknit very vividly when recounting, for instance, her participation in a knitting circle in San Francisco during the time when *Ravelry* went online. The circle was composed like others of "mostly young, white, college-educated women" (ibid., 61) and stands for an essential link between knitting and social media and communication technologies, enabled by the internet, or rather the interknit. Rosner explored the peculiar entanglement of craft and digital practice when creating Spyn, a device that supposedly stipulates the merge of handicraft practice and digital practice even further. Using Spyn, as Rosner (ibid., 63) recounts, knitters produced a "finished artifact [which] worked as a kind of digital container, allowing people to save and retrieve media collected while knitting." This centrally encompassed the "sharing [of] mundane moments" (ibid., 67) and, thus, expand the knitted artefact through the possibility of digitally storing the everyday stories which also contributed to the structure-forming practice of hand knitting. Basically, Spyn worked with an infrared vision recognition to connect media with locations on the knitted artefact. In this regard, Spyn appeared to be an ideal device also for sharing techniques embedded in mundane stories across online communities, such as the emerging knitting circles.

However, even though the device appeared to work just as intended, this is not the end of the Spyn story: Rosner (ibid., 67) continues to explain that as Spyn gained popularity, she was approached by a member of a local knitting guild who asked her "to present the project to her group." In short, becoming a member of this guild herself, Rosner discovers that the knitting practice in this circle is not organised around the interknit, but rather mostly offline and the knowledge, but also techniques, potentially equally embedded in mundane stories, are shared through meetings. In contrast to the young, middle-class, and mostly White knitting circle she had joined before, this group involved "practitioners at the margins—elderly knitters" in different "situations and socioeconomic struggles" (ibid., 77). As a central result, Rosner (ibid.) concludes that knitters are not "a united category of practice." She (ibid., 71) further explains, the guild group differentiated from the interknit circle group as the first was also marked by a "resistance to particular digital tools," for instance, Spyn, despite an initial interest. This, importantly, did not mean that the guild was opposing (digital) technology, but rather that the members of the group had a very different approach to technology than the interknit circles. This could involve, for instance, "using traditional craft skills to transform digital technology" (ibid., 74). Rosner (ibid., 75) further notes: "Computational systems were off-putting when introduced by outsiders as superficial augmentations but also intriguing when incorporated by themselves as enhancements to existing traditions of practice."

What intrigues me the most about Rosner's design storytelling with Spyn is her insight into the simultaneity of dissimilar craft practices across and within different hand knitting circles, while their dissimilarity is interconnected to each group's take on the relation between handicraft and digital practices. Here, the main research interest transforms from how knitting can be aligned with the digital to how to take into account the ways in which needlework is a resource for probing, challenging, and potentially also improving digital practice. Consequentially, this encompasses "formulating engineering *as* craftwork" (ibid., 74; emphasis in original). Working with this inference between knitting and digital practice to emerge and exist is a matter of care, or in Rosner's (2018) words, a matter of *critical fabulation*, while allowing interferences to take shape is an essential part of Rosner's account of critical fabulation as a diffractive design research methodology.

Against the backdrop of my own experience in both knitting and field research in contemporary humanoid robotics, developing a curiosity about the potentials of collaborative knitting between humans and robots that was not random in its choice of research objects, namely the cobot and hand knitting, was pivotal for robotic knitting. Bringing yarn to the robotics lab is more than what appears at first sight as an amusing and playful, but nevertheless interventionist, endeavour. It is also not limited to challenging contemporary automation technologies through the implementation of an unconventional task of HRC. The structure-forming logic of 0 and 1 entails the potential to subvert order and for creative moments of innovating. Hence, knitting practices appear to exceed the idea of a fixed sequence of definable, rather technical movements. Moreover, built on the binary logic of k and p, knitting cannot be thought of as opposed to or prior to digital practice. Equally, they rely on tactile knowledge and the materials' qualities, such as stubbornness.

Resuming here, my engagement with a first attempt of generating data on hand knitting in order to transfer that onto a robot arm lead me to not only revisit, but also to diffract my previously unquestioned assumptions about the relation between the yarn, my hands, the practice of knitting, datafication, the gripper, automation and digital technologies. I did so by engaging in string figuring with a selection of historical as well as contemporary av-

Figure 6: Playing cat's cradle in the lab

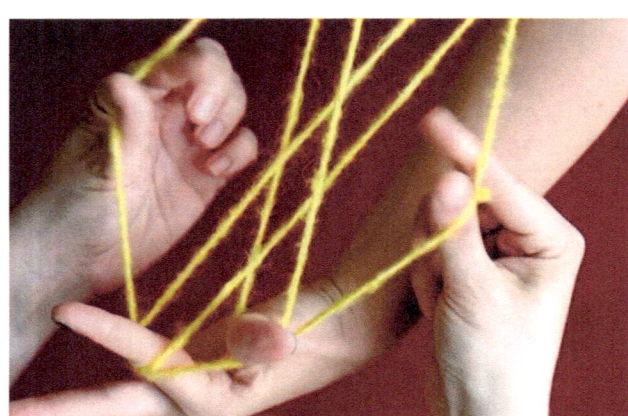

enues of challenging dichotomous relations between handicraft and technology, matter and meaning, as well as knitting and digital practice, but also foundational sociocultural ordering operations like that between male and female, and going as far as the relation between structure-forming as producing order and subversion.

Already the very first practice of knitting in the robotic lab propelled me to develop robotic knitting as a technofeminist tool for not only re-crafting collaboration with robot technologies in a hands-on manner, but also for re-crafting contemporary robotic cultures of striving for robots as co-workers. Notably, it was the initial attempt at turning knitting into data as well as the concomitant enactment of this data which lead me to contemplate hand knitting with a cobot through different eyes, making me wonder how robotic knitting is more than a simple transference of movements onto a robot, but rather the beginning of an exploration of practices of knowing and being in the cobotic lab through forming new stitches with yarn.

2.2 String 2: The Knitting Hands and the Knitting Grippers

In what follows, I will reconstruct selected sociomaterial circumstances of realising collaborative knitting in the robotic lab in detail in order to map this realisation along complex negotiations between different activities and actors—processes which sometimes took different routes than expected, but also showed the potential of the yarn as a material device for literally crossing disciplinary boundaries.

Figure 7: From left to right: Anne Jellinghaus, Melanie Irrgang, Philipp Graf, Raphael Deimel, Pat Treusch, PANDA, and Jan Martin

The project *Do Robots Dream of Knitting (DRDK)* brought persons with disciplinarily diverse backgrounds together, involving feminist science and technology studies (FSTS), computer science, robotics, electrical engineering, psychology, and sociology. From the very beginning, the idea was to establish a project practice which is able to overcome disciplinary boundaries—a practice that I understand in terms of situated co-engineering. Key to this was differentiating between the goal of realising a certain robot task in a robust manner and the path of reaching this as an at least equally central goal of the project. This differentiation allowed us to tackle robotic knitting as more than a challenge for which we had to find a solution, but rather as a method of assembling matters of care in envisioning technological futures with cobots. The process of realising knitting with a robot as the re-junction of digital and craft practice, and its myriad mobilisations and intra-actions of things, ac-

Figure 8: From left to right: PANDA, Anne Jellinghaus, and Katrin M. Kämpf

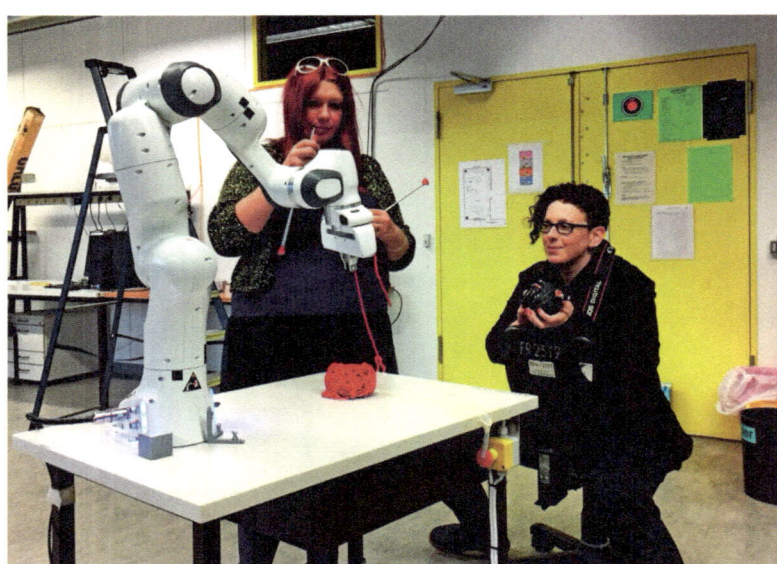

tors, and activities, is as important as the actual act of knitting with the robot itself.

I experienced the constellation at the lab as an extraordinary combination of sociomaterial circumstances, allowing me not only to become a queering witness to contemporary cobotics, but also to become a robotics practitioner, performatively engaging with yarn in this lab as a technofeminist intervention of spinning new threads of human-machine collaboration and imagining cobotic futures. This was made possible due to my access to the robotic lab, the availability of robots, the funding by the Volkswagen Foundation, and the support of the head of the MTIengAge group, Raphael Deimel, and his team—especially Philipp Graf (sociology) and Jan Martin (electrical engineering). Moreover, I was able to hire three additional co-workers who, with their previous training, were ideally placed to help me in bridging the gap between computer science and technofeminism: Melanie Irrgang (computer science), Anne Jellinghaus (psychology & computer science), and Katrin M. Kämpf (FSTS). My team and I became an intrinsic part of the robotic lab—not only through our presence there, but also, and importantly, through weaving our everyday practices of engineering into the fabric of the lab. The lab provided enough space for several persons to work on different projects at the same time. The room is quite large, extending itself over approximately 90 square meters with panorama windows on two sides. The room is divided into two areas. In the larger area, roughly seven to nine computer workplaces are arranged together with four PANDA robot arms, alongside other robotic projects. The space is populated by a varying number of persons, including employees of the MTIengAge group and students working on student projects or working at these computers with robots.

Located on the second floor of a larger university building, the sun would shine through the panorama windows almost year-round. In order to create steady lighting conditions and to keep the room from being heated by the sun, the blinds are almost always closed.

Figure 9: Overview of the main lab space with robots, computer working stations, and team members

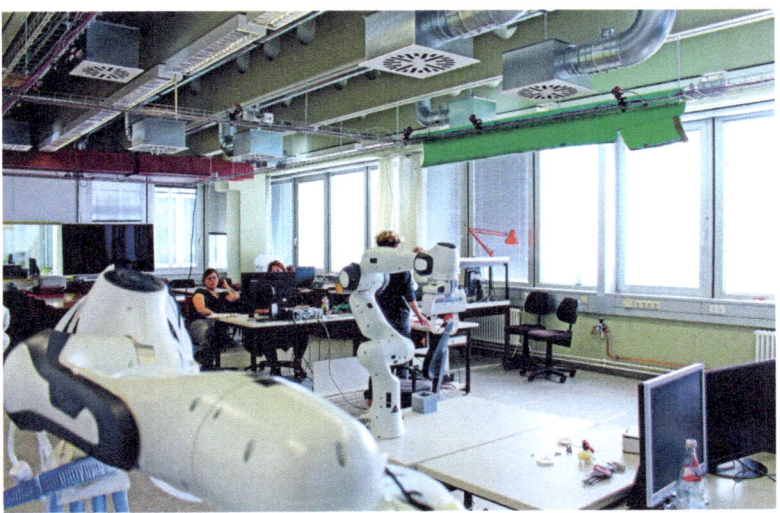

The operation of the computers, robots, and other technical equipment all produces waste heat and I experienced the room's temperature as always a couple degrees too warm, almost unbearable during warm summer days. To air out the room became an important practice and condition for being able to work in and around the cobots for a complete workday, and became the point at which we would pause, take a breath, and reflect on our project work from different angles, allowing us to knit further on our two-sided goal.

Moreover, hand knitting is an excellent example for the meaning of observing and experiencing the working hands in order to learn how to knit. In my case, and as I described above, I used online video tutorials in order to learn how to knit, while in the robotic lab, we worked in teams of at least one experienced hand knitter with one beginner, showing how to knit and then giving instructions during first attempts. We restricted ourselves to teaching and learning one stitch, namely the garter stitch: *k1* last st. This might sound easy, but learning to form new stitches as uniformly as possible was quite demanding even when shown through the hands of an experienced knitter and accompanied by explanations. Mostly during our hand knitting sessions, more than two hands were involved in handling the two needles, testing how hands have to move in order to master the needles, while at the same time, the working hands transferred knowledge and became a central factor of grasping the task of knitting.

More generally, the hand is of central meaning for the development of cognitive capacities, which has been a topic of biology, neuroscience, educational science, but also robotics and cultural science. The hands are used to learn how to count and how to read—a process of incorporating (abstract) signs: the numbers and text (see Robben 2012, 24). In line with recent accounts of embodied human-machine interaction (for example: Dourish 2001), the hand can be regarded as key for embodied experience (see Robben 2012), namely that of *grasping*.[10] This figure then propels an intervention into the Cartesian split between body and mind and the dualism between the abstract, cognitive and the concrete, perceptual and sensuous. The following modification of what can be presumably called Descartes' most popular sentence: "I grasp, therefore I am" (Reiche 2001, 3; translation: PT), depicts the inventive potential of this figure. As put by Bernard Robben (2012, 28):[11] "The hand plays a

---

10   The emerging paradigm of experience design draws on the German term of "begreifen" in order to signify the inseparability of grasping (greifen) and comprehending (begreifen), see: Robben and Schelhowe (2012). The ambiguity of *be-greifen* is similar to the English grasping something in difference to grasping with something.

11   Selected quotations from Robben's (2012) article *Die Bedeutung der Körperlichkeit für begreifbare Interaktion mit dem Computer*, published in German, have been translated to English by Pat Treusch.

special role in grasping (*be-greifen*); it is at the same time a sensory organ—of grasping with, touching and feeling—[and] an organ of action—of grasping, grabbing and manipulating." The cognitive process of comprehending then is based on the hand as that which mediates and translates between the tactile sense and the sense of sight (ibid., 29).

Insights into the relevancy of the hand for human development of cognitive capacities on both the evolutionary and the individual level also give textile craft a new meaning. As Sybille Wiescholek (2019, 112) underlines, "As a structuring element which is produced by hand and with tools, [the textile] is the epitome of an intelligent use of the hand." Such an intelligent use of the hand is regarded as equal to the ability to speak and thus the former is considered to be beneficial for the development of cognitive capacities in humans to the same extent as the latter. In this regard, craft as an activity of using one's hands in an intelligent manner, made me wonder about the possibilities of inviting the robot to become a collaborative knitter with us in terms of an expansion of the idea of distributed learning—in this case from each other's hands, including PANDA in a co-learning experience of the working, knitting hands and gripper.

Assumingly, the meaning of the hand as an indicator and enabler of cognitive capacities and behaviours appeals to AI research, robotics especially. Marvin Minsky's famous *Minsky Arm* from the late 1960ies is prominently placed in the MIT Robotics Collection exhibition of the *MIT Museum* in Boston.[12] Minsky's arm is special as its task to build with children's blocks was realised on this arm without pre-conceptual foundations. Rather, the arm in operation is regarded as the inspiration for Minsky's influential work on a theory of mind. The arm, thus, "gave rise to Minsky's theory that the mind is composed of a multitude of little processes called 'agents'."[13] In this sense, Minsky worked with the mechanics of an artificial arm to derive a theory of mind. Beyond Minsky's approach, the automation of the arm is a core figure of historical prevalence. This is evidenced in mechanical machines as early as Leonardo da Vinci's first robot with arms from 1478 (Moran 2007, 104) to the famous automata heralding the first Industrial Revolution, like Jacques de Vaucanson's flute player from 1738 (ibid., 105), to the first commercial industrial robotic arm in 1962 by Unimate, which weighed two tons (ibid., 108). In comparison, the robot arm PANDA weighs 25 kilograms. It can be

---

12   https://mitmuseum.mit.edu/collection/technology
13   http://museum.mit.edu/150/9

argued that the historical and ongoing fascination with the robot arm unites various interests: a curiosity for the interrelation between body and mind, the automation of tasks previously exclusively accomplished by humans, and the possibilities for improved collaboration between humans and robots.

Sub-String 1: Cross-Familiarisation

At the beginning of the project, it was not only persons who had to get to know each other, but we had to get to know our varying expectations and ideas and the centralised objects of the project, including the knitting needles as much as the cobots. We would mostly begin our workday together in the lab with a coffee meeting where we reported to each other, assembled ideas, and made plans for the day. However, these plans could then be torpedoed by the unexpected behaviour of one of our objects when working with them, like the too-slippery metallic knitting needles or a problem with the cobot's software—experienced on a daily basis. What became clear from the very start of the project was that robotic knitting relies on the collaboration between persons and things, involving a regular, work-structuring exchange, mostly in the mornings, during lunch breaks, and an additional coffee break in the afternoons. In general, our take on the interdisciplinary practice of engineering which is at the heart of collaborative knitting, required a lot of communication across disciplinary knowledge in order to avoid perpetuating taken-for-granted approaches and to avoid compartmentalising the realisation of robotic knitting. The latter would encompass dividing tasks along long-standing disciplinary responsibilities, and thereby might reinforce boundaries instead of overcoming them.

During every full workday at the lab, sooner or later, a point was reached where one of us was asking for fresh air in the room—a point at which we would open the windows, possibly use this pause to take a coffee break, and to reconsider the day's accomplishments from all perspectives involved. Thus, from our need for fresh air emerged an impulse to constantly weave a practice of critical engagement into our own practice of realising the interface between human and cobot through knitting. Another fundamental part of such a practice of critical engagement was a discussion of how to define the outline of DRDK as a technofeminist intervention into reductive, solutionist approaches. This, however, was not an easily completed task, but rather a continuous one which had to be tackled through ongoing conversations and negotiating between the different goals of researching the process of realising collaborative knitting and executing collaborative knitting. While communication was of central relevance to this process, equally important was that we were all meeting in the robotic lab, becoming familiar with the knitting needles and the cobot, and how to operate these objects in order to find ways to bring them together. I suggest understanding this process as one of cross-familiarisation that works with and through the knitting needles, the wool, the cobot in its composition of hard- and software, and the persons, as well as the spatio-temporal arrangement of the lab.

As detailed above, I brought my knitting experience with me to the lab, but was fairly new to handling a cobot. PANDA is a light-weight robot, introduced by its producer, the Bavarian company FRANKA EMIKA, as a robot with "soft-robot performance, smart and industry-ready. Enabling automation for anyone, anywhere."[14] When I encountered PANDA for the first time, I noticed that this robot is endowed with seven degrees of freedom or joints, an end manipulator in the form of a 'two-finger' gripper and LEDs at its platform. This robot is delivered almost completely set-up in a huge cardboard box; one only has to mount the gripper head onto the arm and the arm onto a working station in order to start using the robot.

The MTIengAge lab used smaller tables on wheels as workstations, which reminded me of trolley-tables. On the table's lower tray was enough space for the robot's computer and a multiple-socket outlet. Each robot was mounted to the tabletop with four large screws. An emergency stop was attached to every table—in addition to the robot's own emergency button. The PANDA on wheels turned out to be very practical during our work with it, as it ensured its mobility. We could wheel it over to any computer workplace and work with the robot at this spot.

Upon boot-up, the robot calibrates its joints before going into its initial position, while connecting to an external computer. On this computer, we would start the *General User Interface* (GUI), called FRANKA Desk in order to operate PANDA. Watching the robot's joints calibrating for the first time, the joints making a clacking sound, and the arm moving in a very smooth manner, was very impressive. This already gave me an insight into how the robot can behave as a cobot. At the same time, I also realised that I was not able to estimate the full range of motion, yet, nor anticipate its movements. This lack of knowledge made me uncomfortable to a certain degree. The robot's emergency button, which is connected to the robot by a long cable, helped me

---

14   https://www.franka.de/

Figures 10 & 11: Unboxing & Assembling PANDA

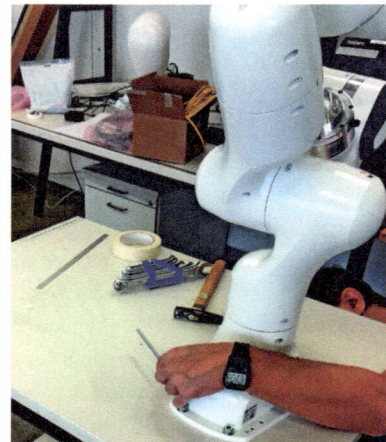

in feeling safe around the robot. One of us always had to solely operate this button to be able to—in case of any doubt—press it and thus avoid possible collision or harm. Nevertheless, it was equally important for me to keep a certain distance to the robot at first.

At the same time, I brought a selection of different needles and wool with me to the lab. As underlined already in the previous chapter, these objects also appeared to be strange and rather improper objects in the context of the robotic lab, causing amusement throughout the duration of DRDK. However, as Melanie herself is quite an ambitious and experienced knitter, and Anne also knew how to knit but needed a refreshment of her knitting practice, we started our project in the robotic lab by hosting several knitting sessions.

Philipp joined us, while it turned out that Raphael also had a robust knowledge of textile techniques, including knitting and crocheting. It became vital to acquire at least a basic understanding of how the needles and yarn have to behave to produce a new stitch, in order to generate ideas and engineer solutions for transferring this movement to our robot arm. At the

Figure 12: Knitting in the robotics' lab

same time, Melanie started the main blog entry series *Knitting for Computer Scientists and Engineers* on our project blog.[15]

Given this initial situation of a cross-familiarisation between multiple agents, objects, and knowledges, I suggested to work with three different scenarios of collaborative knitting with PANDA, with the plan to realise all of them within the first months of the project to become more familiar with each other and with each object. These realisations should serve as a testbed for evaluating our approach towards collaborative knitting, but also for developing an account of interdisciplinary, situated co-engineering between FSTS and robotics in concept and practice. The three scenarios of collaborative knitting between the robot arm PANDA and 'us humans' were characterised by different degrees of physical distance to the robot arm as well as varying degrees of collaboration as depicted in the following.

In Scenario 1, a knitter is in need of a kind of 'third hand' which is helping in unravelling further the ball of yarn during the knitting of rows. The flow and tension of yarn is a crucial factor in knitting an evenly structured artefact. Without such a 'third hand', one might have to put down both needles in order

---

15   https://blogs.tu-berlin.de/zifg_stricken-mit-robotern/

Figures 13, 14 & 15: The three scenarios of collaborative knitting

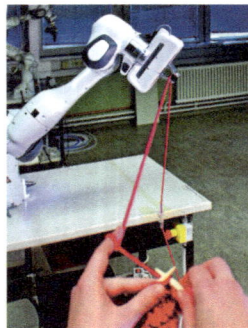

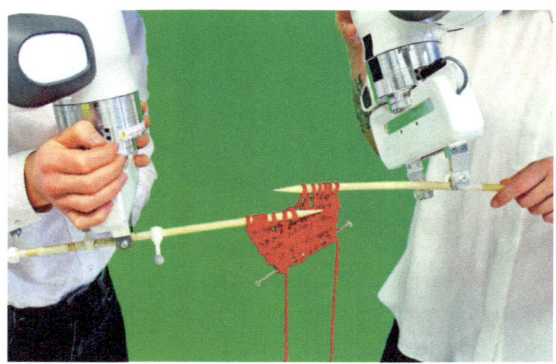

to grasp the ball of yarn and unravel it further. Thus, the cobot's task in this scenario is to unravel the yarn when needed. Clearly, in this first scenario, one of us could knit like they would normally do and remain in a safe distance to the moving PANDA. The degree of collaboration is still quite low. Thus, this appeared to be an ideal initial scenario for getting to know the robot more, watching how it moves through space and becoming gradually more familiar with its capacities as a cobot.

Scenario 2 then was supposed to be based on the initial experience with the cobot from the first scenario. In this scenario, one of us was already knitting with PANDA, handing over the empty right needle to the cobot, while still operating the left needle with the knitted piece and the last knitted row. This clearly requires a fine-grained collaboration between the robot and the human as the robot has to perform the following movements: pierce through a stitch on the left needle, pick up the thread through the stitch (forming a new stitch) and finally let the 'old' stitch slip from the left needle. Informed by my first attempt at a datafication of my own hand knitting, I knew that knitting requires a high degree of orchestrating between the movements of the right and the left needle. So, in this scenario, controlling these two needles is a task which is divided between the person and the robot. The challenge then is not only to realise the 'right behaviour' in the robot arm, but also, and importantly, to coordinate these movements with the movements of the other needle, and to realise this in a collaborative manner.

Scenario 3 is built on the second scenario. After the cobot has been trained to take over the right needle, the next step would be to train a second robot to take over the left needle so that the two robot arms would be knitting together, while the role of the person would be to give and guide the thread. In this sense, the person would have to become the sensory guide to the two robot arms, enabling a form of collaboration in which they become 'the eyes and ears' of the robot—at least to a certain degree. This, I imagined, would display the most advanced form of becoming familiar with each other. The plan was to move along these scenarios, working as a group on realising each scenario and tracing the enmeshment of actors and objects, as well as the emerging negotiations, challenges, and solutions, from the perspective of the queering witness who is herself immersed in these engineering practices.

Sub-String 2: *k1* last st—First Collaborative Rows

When starting to work with PANDA, we first realised individual movements as well as shorter sequences of movements on the robot arm, using the GUI FRANKA Desk, and by assembling ideas, materials and practices of realising collaboration. In what follows, I will trace the implementation of the scenarios of collaborative knitting through selected steps of robotic knitting.

In accordance with the paradigm of visual programming, FRANKA Desk has different color-coded buttons. After opening the program FRANKA Desk, the first step is to choose between the Cartesian motion and the joint motion apps. The first enables the execution of a movement based on remembering the position of the end manipulator (the gripper) on the x-, y-, and z-axes at selected points. Working with the second app, the robot arm will remember the constellation of every joint at each point, in addition to remembering the position of the end manipulator. The location of the end-effector on the x-, y-, and z-axes and the concomitant joint configurations are called *states* in robotics and the model of computation behind this account of states is that of a *discrete* or *finite state machine*. In short, states are categorised into either a start state or an input-related state, amounting to a movement trajectory while the state machine regulates the transition between individual states. The main difference between the joint motion and the Cartesian motion apps is the transference from one state to another. To memorise points and the joint configuration is called *forward-kinematics* and, akin to that, *inverse-kinematics* is to memorise the location in space and compute the joint configuration needed to get there (Irrgang 2019). As Irrgang (ibid., 1) explains further in her blog entry on motion planning: "Using a joint motion App, every state's joint configuration is remembered for later playback. Using a Cartesian motion App, only position and orientation in space are remembered, and inverse-kinematics is applied for every state of the movement trajectory." An initial step in realising collaborative knitting was to test the differences in working with a state machine based on either a forward-kinematic model or an inverse-kinematic model. When working with the Cartesian motion app, we would sometimes become very puzzled by the results the inverse-kinematics produced, making it harder for us to grasp the execution of movements, especially given that knitting movements are very small motions. It also happened sometimes that the computed joint configuration would trigger the cobot's collision safety mechanism, provoking a lockdown of the cobot. Thus, we decided to work mainly with the joint motion app, enabling a forward-kinematics model which made the transitions from state to state very tangible to us.

According to our plans, the first scenario we realised was Scenario 1 in which PANDA gives thread while a person knits. The aim of this scenario was for the persons collaborating with PANDA to start working with wool, while remaining at a large enough distance to the arm to not stand in the way of its range of movements. The idea was that the gripper pulls the thread to unravel it from the ball of yarn. However, the first thing we noticed was that the gripper was unable to grip yarn-like things. The yarn would always slip through the closed end effector. The second challenge we encountered was the movement of the ball of yarn which showed an uncontrollable and unpredictable behaviour when being pulled. Together, we came up with a solution for both challenges. We built a guide rail for the yarn so that the position of the thread was allocatable for us, and we could position the gripper in relation to the thread. For this rail, we used a shorter piece of a pneumatic hose, which we mounted onto a smaller piece of wood, which we then attached to PANDA's table. Next, we came up with the idea to not have the gripper properly grip the yarn, but to train a movement in which the arm moves in such a way that the yarn is wrapped around the gripper, and through movement the thread would be unravelled from the ball of yarn.

Figures 16 & 17: Realising Scenario 1

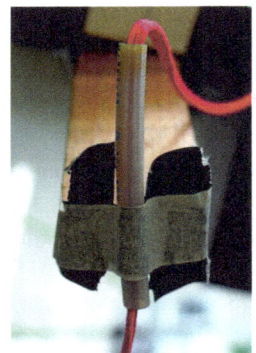 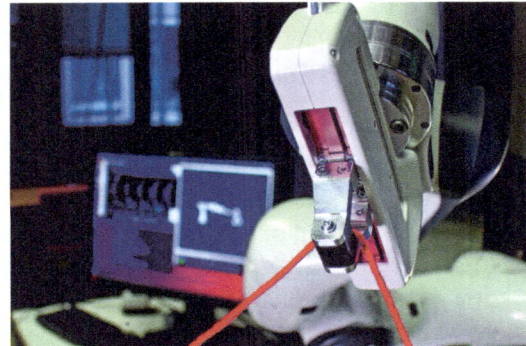

Scenario 1 was a very nice experience of learning how to operate PANDA, that is, learning how to implement a movement and learning how the arm behaves when executing this movement, but was also a nice experience of defining and solving challenges in realising our planned scenarios of collaborative knitting with the robot from an integrative perspective. However, the degree of collaboration between the knitter and PANDA is rather low in this scenario. Hence, we quickly moved on to Scenario 2, in which PANDA is supposed to operate the right needle. We began by asking what this actually implied. As by now everyone was familiar with hand knitting—at least on a basic level—we started by thinking about the whole idea of collaborative knitting in a very hands-on manner.

Usually, when I am knitting, it would not occur to me to hand over the right needle to someone else in order to knit with another person, but if I should, I would probably talk to this person, show them how to hold the needle, and constantly communicate with this person. Ideally, the person who is knitting with me, already has a concept of hand knitting and thus understands not only what the right needle is supposed to do, but also how the two needles will have to interact. However, when aiming to realise this with PANDA, we had to acknowledge that the robot has neither 'eyes' nor 'ears', that is, it lacks the sensors to communicate in a somewhat human-like manner, nor does it know that we are knitting or that there exists a thread that it has to master with a needle in the first place. PANDA will not have a concept of hand knitting. However, at the same time, PANDA belongs to the newest generation of flexible light-weight robot arms, designed to take on new automation challenges, including a host of scenarios in which PANDA becomes a kind of a helping hand to 'us humans'.

I understand the figure of the helping hand as a diffraction pattern emerging from my string figuring with PANDA, wool, knitting needles, technofeminism, and disciplinary boundaries. Already during my doctoral thesis and throughout my postdoctoral work, I noted that robotic hands in their signification of tactile capacities play a major role in imagining, building, and representing robots that are geared at varying forms of so-called 'social interaction' that entail acting with humans in close cooperation. This especially appears to be a foundational aspect in re-locating the formerly industrial high-end functional robot to the realms of the service sector or the household. Clearly, core capacities that are regarded as essential and useful, for instance, in the automobile industry, like the ability to lift heavy things, and therefore to outmatch humans in their physical strength, no longer appear attractive in a setting of human-robot-proximity (see Treusch 2015, 78). Imagine how a person feels when the metal robotic gripper, which appears indestructible and is the epitome of the robot's force, is reaching for the person's hand in a situation like handshaking. Here, the gripper has to be endowed with actuators, as well as sensory and control capacities, which allow it to close itself around a human hand with the right pressure and velocity—not only for a human to not fear the danger that the robot might squash their hand physically, but also to perform the movement of shaking-hands in a manner in which the movement becomes legible as a shake of hands and is therefore socially meaningful. Thus, the robot has to not only be able to grip, but also to grip in the manner of handshaking, which describes a sociomaterial realisation of human-machine relating. Thus, the hand as that which can create, caress, and destroy appears to be of significant symbolic relevance for imagining and realising present and future scenarios of HRI. Robotic knitting not only accommodates this significance of the hand for developing cobotic futures, but also becomes an agent in creating scenarios of how human and robotic hands and grippers can work together. Further, robotic knitting in its technofeminist, interventionist stance aims at tweaking the idea of the capable hands as the embodiment of a friendly encounter by reading it against the grain, or rather, by queering the implied usefulness through a collaborative hand knitting practice. In addition, handling needles and yarn together is a practice which evokes the (historical) junction between computer technologies and textile craft as well as a practice of opening up a field of tension between *techne* and *logos*.

Contemplating about the working hands here further, I become increasingly interested in the nature of hand knitting as a practice. Hand knitting movements are characterised by non-uniformity instead of uniformity. The MoCap picture illustrating this (see *Figure* 3) attests to the complexity of hand knitting as a technique which requires training and experience—even on the level of the most basic stitch, the garter stitch—in order to become able to continuously form new stitches and produce a knitted piece with stitches as even as possible. The latter involves the ability to constantly adapt to the material obstinacy of the yarn. To become proficient in hand knitting is achieved by embodied learning in which hand and mind are working together in using the needles to realise a textile structure. Moreover, what I want to emphasise here is the collaboration between hands, needles, yarn, and concepts as foundational for the creation of textile structures. Thus, hand knitting can be regarded as provoking a junction between the working hands and gripper, which opens up possibilities for re-designing human-machine-relations along knitting practices of collaboration.

Tim Ingold helps me in understanding hand knitting as a process of creation which speaks well to the technofeminist, performative and diffractive outline of this project. Part of Ingold's (2009, 92) rich oeuvre is to overcome a "hylomorphic model of creation" in favour of what could be called a process-ontological model. This model "assigns primacy to the process of formation as against their final products, and to the flows and transformations of materials as against states of matters" (ibid.). Drawing on Ingold's notion of creation, it is exactly the myriad unfoldings and becomings of a form which build the core of practices of making; to adhere to and engage in these forces and flows then are key dimensions in learning a *skill* like knitting. Ingold (ibid., emphasis in original) further suggests calling this "the *textility* of making" or the textility model in contrast to the hylomorphic model. Knitting encompasses to follow the lead of the yarn, how it behaves, but also the lighting conditions and colour of the yarn, its material qualities, like how firm individual threads are spun, the slipperiness of a needle (for instance that of a bamboo needle compared to that of an aluminium needle), and the emerging stitches of a row. Foregrounding material agency over the control of matter through the activity of human hands here, I draw further on Ingold's work on textility to conceive of hand knitting in his words. Making a stitch in line with Ingold (ibid., 94) "is not so much imposing form on matter as bringing together diverse materials and combining or redirecting their flow in the anticipation of what might emerge." The working hands of knitting are but one factor in this process of anticipation and combination, equally *"possessed by the action"* (ibid; 95, emphasis in original) as are the wool, the needles, and the technique (the garter stitch). This account of hand kitting tweaks the meaning of the working hands and gripper across the assembled fields of knowledge and practice in this project, making them a matter not only of embodied inter- as intra-action, but also of a re-valuing of forming over forms, materials over things, in human-machine interaction, and opening up the possibility to re-craft collaboration across the different agents, materials, and flows engaged in the task of hand knitting collaboratively. Describing the working hands in our hand knitting sessions reveals how this is already the case in human-human practices of teaching each other how to knit, which I experienced as very tangible and graspable processes of transferring skills of combining and following the materials of hand knitting in order to be able to create an artefact. In addition, learning how to operate PANDA can also be captured as a sensuous, embodied process. At stake is an exploration of the textility of knitting as a skill, in relation to human-machine collaboration through knitting.

When starting to realise Scenario 2, we decided to work mostly as a team of three (Melanie, Anne and me) when realising a task with the robot: One of us sitting at the computer, the other standing next to the robot, and the third person holding the emergency button (see *Figure* 18).

Right at the beginning when we started to knit, one foundational question emerged through our haptic engagement with the selection of needles and wool which I initially brought with me: What material qualities must the yarn and needle have to enable human-robot collaboration through knitting? Clearly, we had to investigate the question of knitting materials further.

We quickly found out that each of us might answer this question differently when it comes to our personal experiences and preferences. Depending on how tightly or how loosely one knits, the gauge, firmness, and the twisting of the yarn matter a lot in the forming of new stitches. In addition, the

Figure 18: Working on Scenario 2

Figures 19 & 20: Needles and yarn

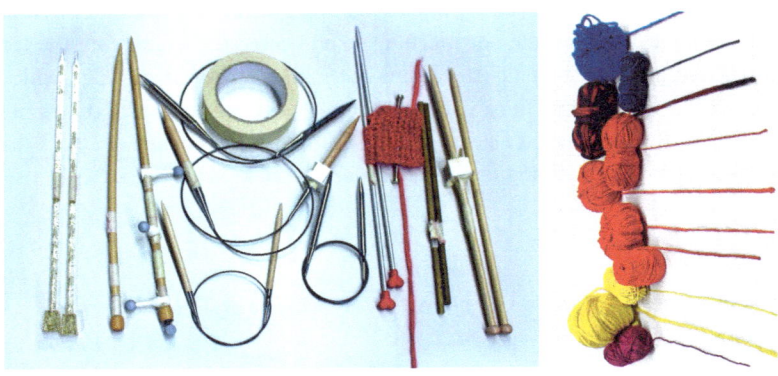

materials of the needles shape the behaviour of each yarn differently. Our experience was that bamboo needles show the right kind of slippery qualities between too slippery and too dull for learning to knit and for teaching both other persons and PANDA to knit. Our next insight then was that the needles should not be too small and the gauge of the yarn not too big. Such a combination of rather big needles and a not too thick yarn produces larger stitches and makes it easier to pierce through the stitch on the left needle with the right needle to then pull the yarn through the old stitch and form a new one. The yarn which became our preferred one is a quite soft quality yarn which is not twisted in a classical manner, but woven. This microstructure made it at the same time flexible and firm enough so that, for instance, it does not dissolve into individual threads when the yarn is moved around a lot during the knitting—a behaviour that other classically twisted yarns have shown.

While finding the right needles and yarn for us, we also discovered that the robot's gripper is not only not made for gripping yarn, but is also not made for gripping needles. Rather, the core grasping target for the gripper seems to be box-shaped things. The inside of the two gripper 'fingers' is endowed with plastic grooves. These are clearly meant to enhance the gripper's ability to grip. However, in the case of working with knitting needles, the needles could not be gripped properly due to the grooves, so the idea behind their design turns out useless for our desired application of the gripper.

This mismatch between gripper and needles needed to be fixed. We started to tinker with different materials and practices. First, we used crepe tape to increase the gauge of the needles. This already worked pretty well towards fixing the mismatch between needles and grippers. However, the relation between needles and grippers was still far from ideal. There was also a 3D printer in the lab and one morning when I entered the lab, Anne and Jan surprised me with a small 3D-printed box with a hole in it, through which the needle could pass. Thus, with the help of the box, we turned the round needle into a box-shaped object for PANDA to grip. This turned out to be the ideal solution for enabling the gripper to grasp the knitting needle.

Figure 21: The 3D-printed box; Figure 22: The box with the needle in the gripper's grasp

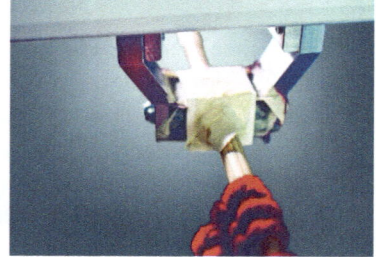

Notably, assembling the materials needed to realise Scenario 2 was not a linear process as it might appear when recounting this here. Rather, we would spend a large amount of time tinkering and testing in a rather playful manner of engaging with materials. This would include everything that was already part of the lab (like the pneumatic cable we used in Scenario 1 or the 3D printer), materials we newly introduced to the lab (like the wool and knitting needles), as well as constant conversations across disciplinary boundaries, which made the whole process a team effort that unfolded through iterative circles of engaging with materials, defining challenges, discussing new approaches, testing ideas, and finding workable solutions. However, we also worked with a provisional account of solutions that was open to revisions and changes. This openness proved central to establishing a practice of following the flows and probing combinations of materials. Furthermore, my account of realising Scenario 2, namely to give the right needle to PANDA, makes tangible the ways in which this is not one discrete task, but rather is composed of several sub-tasks involving various steps that we had to work through in order to accomplish the task. Describing our sub-tasks, and the ideas and practices involved in detail here, I want to make tangible the ways in which robotic knitting displays what Ingold calls the textility model of creation, making it necessary to follow the flow and composition of materials and practices.

Another integral part of realising Scenario 2 was to understand the operations of the right needle. Forming a new stitch without the involvement of visual or tactile sensors as we, the humans in this project, were endowed with, is obviously a different task. Thus, we decided to work with the differentiation of basically three individual movements of the right needle as they are explained to persons who are starting to knit. These movements are: (1) stitch through, (2) wait and pull yarn through old stitch, (3) wait and let old stitch slip from left needle. We could then define each movement as a set of points in space located through the end-effector in combination with a specific configuration of joints and called this task *knit the right needle* (krn). At first sight, to transfer knitting onto the robot arm by defining sub-sets of points in space might appear like robotic knitting is an easily accomplishable task. However, what complicates this is the collaborative dimension of robotic knitting, namely the left needle in the hands of a person meant to be knitting together with the right needle held in the gripper of the cobot. Thus, we not only had to make these three movements transferable, but also had to have the robot execute these movements in a way that they made sense in the overall setting of knitting collaboratively.

Prerequisite for starting to implement krn was to already have knitted a couple of rows, before handing over the right needle. The first few rows of a new knit artefact turned out to be too delicate and fragile to combine with the effort of handing over a needle to the robot. Thus, we would first knit a few rows before handing over the right needle. Working with a knitted piece on the left needle, and now perfectly prepared to hand over the right needle, the first steps with the cobot were easy, encompassing releasing the brakes, enabling the teaching mode, and using the gripper app to tell the end-effector to close around the needle with the white cube. In the next steps, joint motion apps had to be combined with pause apps in order to assemble the order krn. Each joint motion app had to be filled with information. Working at the computer, this combination of apps was just a mouse click on differently color-coded buttons. Melanie who was operating the robot arm in the so-called teaching mode, then had to start with defining points in space while moving the arm to these points. This involved pressing the guiding button above the gripper, which enabled her to move the arm. Notably, the guiding button has to be pressed in the right manner: If pressed too soft, the arm wouldn't move and if pressed too hard, this would cause an emergency stop and the arm again wouldn't move. So, becoming able to guide the robot arm is to learn how to press the guiding button. This can only be learned through a trial and error approach. Then, pressing and holding the guiding button in the right manner, one can guide the robot arm to move in a specific direction. Importantly, the movements we wanted to implement on the robot had to be always perceived of as well as performed in relation to the left needle. In the scenario described here, this meant transferring Melanie's movements of the right needle onto the robot arm, which is far more than 'just' the three basic movements of the garter stitch as identified above. Rather, it also encompassed not only the individual hand knitting style, but more generally, the ways in which the needles have to be orchestrated in order to form new stitches.

Stitches, yarn, and the knitted piece are constantly moving when knitting, and it seems impossible to predict how they will move. Splitting the activity to knit between a person and PANDA not only emphasises how complex knitting is, including the many different smaller movements which are natural to an experienced knitter, but also emphasises the complexity of realising this mode of collaboration between humans and a cobot. Thus, hand knitting is a movement which can be formalised and automated, as the example of the industrialisation of knitting illustrates nicely, and at the same time hand knitting also always exceeds the reduction of this practice of forming new stitches into a set of formalisable movements. It seems like at the core of hand knitting remains a textility of creation which defines this practice as the engagement with flows and combination of materials that cannot be controlled by hands, but rather instructs a process of being possessed by actions in the forming of new stitches. A core challenge was to do justice to this quality of the practice of hand knitting—to make it our matter of care—and thus to transfer not only a set of sub-movements, but rather a *skill* onto the robot—of course within the limits of the technological constraints of the cobot, while acknowledging that the hands and gripper needed to be possessed by action as integral to knit collaboratively.

Against this background, zooming further in onto the practice of transferring knitting movements to the robot, the challenge of transferring more than a movement becomes tangible on many occasions. As I will continue to argue, it is precisely the quality of knitting in its not only structure-forming (order and subversion) quality, but also the working together of the hands and gripper as well as the flow of materials, which also make tangible the many different forms of collaboration that are prerequisite for realising the task of knitting collaboratively. A very vivid example is the practice of guiding, as the core element of PANDA's teaching mode. After learning how to press the guiding button as explained above, the person who is guiding PANDA also had to then not only press the button throughout the entire movement that the robot should be trained, but also first execute the movement and while doing so, second, press the enter button at the robot's pilot in order for it to memorise points on the x-, y- and z-axes. Again, this might appear to be a rather easily accomplishable task, nevertheless, it demands that the person guiding PANDA does different things at the same time, like emulating a desired movement through moving the robot, deciding points in space that the robot should memorise in order to be able to execute this movement afterwards, and pressing two buttons.

Thus, Melanie had to first study herself practicing hand knitting, especially how her right hand with the needle moves, to then become able to abstract individual movements and translate them into several stations of her right needle that can be thought of as points in space—always in relation to her left hand with needle. Hence, we discovered that the practice of guiding PANDA is itself something that we had to practice intensively. Notably, this turned out to be an embodied process, demanding an interplay between materials and agents, but also involving different senses. In line with Ingold, I

Figures 23 & 24: Melanie guiding PANDA

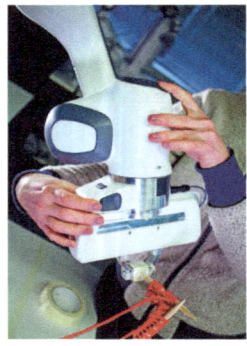

describe this process as stipulating a practice of becoming possessed by actions. Precisely at this point in which guiding PANDA unfolds as a *skill* on its own, the intra-active nature of collaboration between humans and cobots crystalises. When guiding PANDA, bodies intra-act, deployed by what Morana Alač (2009, 496) calls practices of "'getting into' the body of the machine." Getting into the body indicates a set of practices of attaining a multi-modal and embodied sense for the robot. This account of interaction challenges the idea of the individual body and of agency as a property of the individual and thus tweaks interaction towards intra-action: The knitting movements of the right needle of my hand-knitting practice become the guiding of PANDA (conditioned by touching, moving, engaging, experiencing, material flows, but also obstinacies and many other sociomaterial factors) and the subsequent movements of the robot arm, which again become the movements of the right needle of my hand knitting practice. Guiding then is a lot about finding a shared rhythm of moving together. Highlighting the dynamics of bodies in intra-action, I suggest an account of collaboration at the human-cobot interface that takes into account these dynamics as its matters of care.

Moreover, when working in the constellation as shown above (*Figure* 18), we would use the possibility to hit the enter button on the screen instead of on the pilot, so that Melanie did not have to press the enter button any longer, but could transfer that to my fingers on the computer mouse. We agreed that she had to signal me with the word "now" when I had to click on the set a point button on the computer screen. This division supported Melanie in the highly pre-requisite guiding. In addition, my role cannot be reduced to clicking the mouse as neither can Anne's role to identifying a potentially dangerous scene and pressing the emergency stop button. Rather, the three of us were constantly collaborating, Melanie explaining her movements with the robot arm, drawing Anne and me into her practice of getting into the body, making decisions on how to move and where to set a point, possibly erasing a point. Thus, PANDA, Melanie, Anne, and I were acting in concert. Anne and I would eventually sit and stand at our positions while observing Melanie and at the same time, also getting into her/the robot's arm with our own right arms in order to guide with her. We literally and figuratively became co-guiders.

In the next step, we could start testing the actual execution of collaborative hand knitting. The spatial arrangement and composition of persons, computers, and the robot would remain the same for this. My role now was to pull up the order krn, to release the robot's brakes, and to then wait for a sign from Melanie that she is ready before clicking on the play button. At the same time, Melanie would check if PANDA was in the correct start position, and if not, guide it there, then put the right needle in PANDA's gripper, hold the left needle as well as the yarn with her left hand and position herself in a way that made it possible to start knitting now.

Figure 25: krn on FRANKA Desk

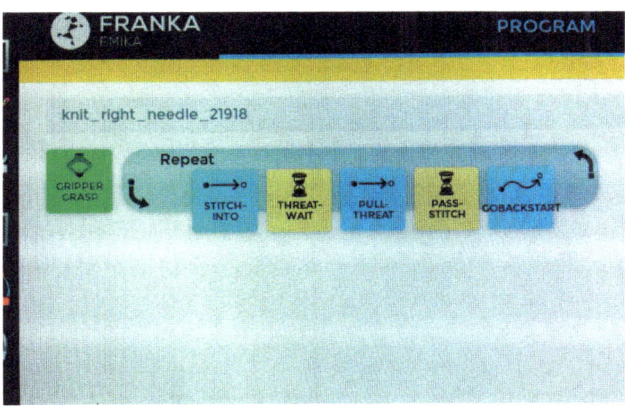

As Anne was continuously watching over this scenario, after the first couple of executions of krn, she figured out that we always needed "five warm-up stitches," before the actual knitting started to be performed more smoothly. This established itself as a rule which all of us were able to confirm through our experiences. Further, I regard the phenomenon of the five warm up stitches as displaying how not only guiding, but also the execution of a movement with the cobot illustrates the intra-active nature of knitting collaboratively, that is, entangling oneself in the assemblage and flows of materials of forming new stitches together with the cobot as a practice that challenges precision and accuracy. Carrying out krn then is not simply the execution of a movement, but rather the experience of this movement through the robot arm, based on the constant orchestration of hands, grippers, needles, buttons that is enabled through practices of getting into the body, the embodied and multimodal process of attaining a sense for the robot—a process which is—equally to guiding—a lot about finding a shared rhythm of moving together.

After successfully implementing collaborative knitting, we also invited others from the MTIengage group to test the realisation of Scenario 2. On one of these occasions, Philipp took over the left needle, who at that time still was not very secure in practicing hand knitting and had not been a knitter before the project. In this composition of persons realising krn, the robot's ability to perform the krn movement with a quite high precision and in endless repetitions worked nicely towards teaching Philipp how to knit. Now, Melanie and I would look both Philipp and PANDA over the shoulder, guiding Philipp in how to get 'into' the arm of the robot with the right needle as his right arm as the robot arm. We, the three observers, in turn, nevertheless had to attain a sense for the robot's arm as Philipps arm, and how we could support the learning process. This was an embodied and situated experience of developing and sharing skills across entities. Thus, the example of learning to knit with PANDA makes tangible the ways in which not only attaining a sense for the robot, the yarn, the needles, and the collaborative knitting are enmeshed practices of relating, but also the concomitant process of assembling matters of care as well as how different agents engage collectively in krn.

Finally, after working on and experiencing Scenario 2 for a while, we decided to move on to Scenario 3, even though, built on our experience, we already assumed before testing Scenario 3 that Scenario 2 might be the most exciting of the three scenarios with regard to probing collaboration between humans and cobots. What appeared very obvious to us was to base this new implementation on what we had learned from the previous implementation of krn. Thus, we decided to combine krn on one PANDA with *knit the left needle* (kln) on a second PANDA. Moreover, I believed that characteristics of the robot arm, like the ability to execute always the same defined movements iteratively in a very precise manner, would be helpful in having two arms working together, while a human then could take over the part of orchestrating the flow of materials in forming new stitches. We worked in a team of more than three persons in order to realise this scenario and quickly found out that the guiding of the right arm and left arm could not be separated, but needed to be reciprocally related: One of us was guiding the left needle and the other the right needle, attempting to train each robot arm the pre-defined movements while also knitting with each other. This turned out to be highly challenging for everyone involved.

In addition, and especially compared to Scenario 2, the collaborative aspects in Scenario 3 appeared to be limited to either human-human or cobot-cobot collaboration, neither of which were my main interest. However, we continued with our attempt, driven by a curiosity about how our efforts will turn out. The same conditions and circumstances applied to the execution of krn in combination with kln as to the execution of krn: The persons at the computer each had to press the play button with a mouse click. What we noticed immediately was the importance of timing in this. Both play buttons on each computer had to be hit at absolutely the same time in order for the two arms to move in synchronicity. We first started to try to count and hit play, but this did not work so well, so Melanie proposed to use the rhythm of a song we all knew and define a point in the song at which we would press the button. This worked much better in reaching the desired synchronicity. However, despite the identical settings on each robot arm and reaching the synchronicity in pressing the play button, the execution of movement of each cobot turned out very asynchronous, so that one needle would operate approximately 10-15 centimetres above the other needle. This happened every time we repeated the execution of kln in combination with krn. As a result, and, at the end of the day, mainly my frustration with Scenario 3 on the level of human-cobot collaboration as well as its execution, led to the decision to not further pursue this idea.

Sub-String 3: Improving Collaboration: From Precision to Increasing Flexibility

When implementing krn, we could adjust the execution of movements by changing the settings, mainly by changing the velocity of it. In the following screenshot (*Figure* 26), each sphere represents a state.

Figure 26: Adjusting settings in the joint motion app

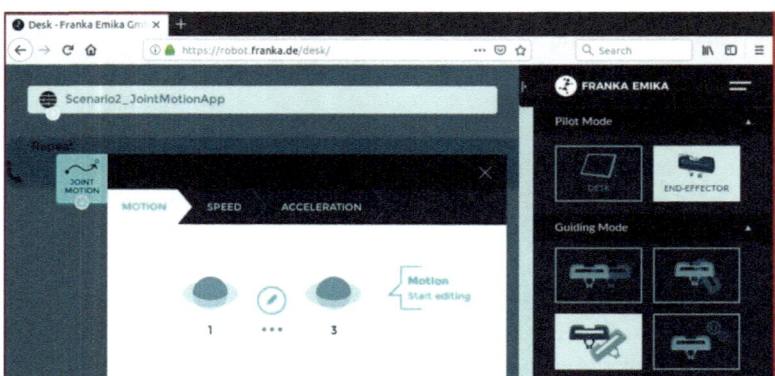

Despite this possibility of adjusting the execution of states, when Melanie trained the krn and I took over from here in the execution of collaborative knitting with the cobot, I would experience different challenges. One challenge was that the way PANDA was moving the right needle represented Melanie's hand knitting practice, which differs from mine. This, again, underlines the ways in which hand knitting does not only consist of a set of sub-movements which are apparently the same for every hand-knitter, but that hand knitting is a skill. Another challenge was the velocity of the execution of movements by the cobot. I preferred a faster execution, for instance, compared to Melanie and Anne. Thus, in our experience, the possibility to change the basic settings of krn was very helpful in adjusting the collaboration between PANDA and different persons through knitting.

Besides these and other rather easily accomplishable technological improvements of collaborative knitting, we also experienced a certain boredom with krn after a while. First, of course, we were all excited and fascinated by our success in implementing this movement onto PANDA in a quite robust manner which allowed each of us to engage in knitting with the cobot, from the re-engineering of the knitting needle by making it a box-shaped object to adjusting the velocity to individual preferences. However, the execution of krn also encompassed certain qualities that we could not change, like the very basic principle of executing a movement along points on the x-, y-, and z-axes on the one hand, and on the other that the cobot arm is rigid during this execution. The way PANDA would execute krn is that the end effector would go to every memorised point, constantly switching between acceleration and stopping, and thus kind of moving in what could be framed as classical robotic movements, characterised by fragmentation in the flow of movements. We were affected by the disruptive quality of both behaviours that cause fragmentation, namely jerkiness and pausing, and while we experienced them as helpful in the beginning of the project, we now started to regard them as limiting our experience of collaborative knitting. Melanie, who knitted the most with PANDA, described it as painful and fatiguing not only to wait for a transfer to be completed, but also to have to work in states, and therefore in fragmented movements, when executing krn. Such a movement from state to state is implemented by the preferred motion planning app, joint motion, and results from the forward-kinematics on which this app is based. What we experienced in an early stage of the project as a benefit of this model, the comprehensibility of states, thus, was also the source for our frustration with the execution of movements at a later stage of the project.

Luckily for us, throughout the duration of the project DRDK, the head of the MTI-engAge lab, Raphael, was also working on his behaviour control program, called *PhaStaProMP* (Deimel 2019a; 2019b) and invited us to try to implement krn not only using FRANKA Desk, but also PhaStaProMP. The major differences between the visual programming of the GUI and the behaviour control program developed by Deimel with regard to our project are first, how one trains a movement and second how the robot then performs the trained movement. In what follows, I will briefly describe the technological differences, then give details on how we approached working with PhaStaProMP in order to make tangible the ways in which the latter was able to improve our practice of collaborative knitting.

The basic idea behind PhaStaProMP is to enable "reactive interaction" (Deimel 2019a, 1) between a robot and a human. Notably, this encompasses reworking the model of the discrete state machine towards a "phase-state machine" (ibid.), which "can implement regular state machine semantics, but it additionally has the built-in capability to provide and adjust phases and blend consecutive movement primitives for smooth operation" (Deimel 2019b, 1). In short, working with the blending of states, Deimel's phase-state machine model PhaStaProMP would ideally erase the fragmentation of movements as it works with "dwelling in *states*" (ibid., emphasis in original) in contrast to the "*state transitions* (guarded jumps)" (ibid., emphasis in original) of the discrete state machine based app we had been working with so far.

Such a dwelling in states is realised through different mathematical methods of implementing transitions between states, including, for instance, working with a differential quotient to establish a pulse for the process of cycling through states, as well as probabilistic decision-making. In short, the outline of the PhaStaProMP state machine is to combine the definition of a state with different circumstances of its execution as well as the synthetisation of that state in order to generate a continuous movement. What intrigued me as a technofeminist scholar, neither trained in robotics nor mathematics, was what I understand as the idea to realise an approach to robotic movements which operates with the basic concept of a state and at the same time challenges the very notion of a state by re-working the relation between states and transgressing the notion of proper boundaries between motions. In my view, this approach to robotic movements can be read as reverberating with the idea of the hand knitting practice in terms of being possessed by action—so, the question at hand now is: How did PhaStaProMP change our krn practice?

Training PANDA to perform a movement with PhaStaProMP is radically different from working with FRANKA Desk, even though we would still be working as a team distributed at the PANDA, emergency button, and computer screen stations. The interface was no longer a GUI but a *Python* command prompt window. Thankfully, Deimel wrote a step-by-step how-to manual of commands and settings for working with PhaStaProMP which was also intelligible across disciplinary diverse backgrounds. Working with PhaStaProMP, the first step was to activate the phase state machine, then to train states through demonstration.[16] The latter is comparable to the guiding process, but not the same. As soon as the guiding button is pressed, the motion will be recorded, including the configurations and velocity and the recording stops when the guiding button is released. The recording of movements produces training data with which it becomes possible to generate a trajectory of defined states. We only needed three iterations to work with an additional number of synthesised movements that were generated based on our iterations. Each demonstration should be as similar as possible, preferably producing variabilities only within a certain degree and therefore

---

16 For a detailed description of this process, including video footage, see Melanie Irrgang's blog entry (2019) on *Time for Action – Scenario 2*, https://blogs.tu-berlin.de/zifg_stricken-mit-robotern/2019/05/03/8-knitting-for-computer-scientists-and-engineers-scenario-2/

deliver the data needed to first simulate and second learn a movement from demonstration. The demonstrated movements have to reveal the range of the ideal movement, also allowing to define that which limits this range, namely the discardable. To produce such data requires a form of what could be called disciplining of one's own body, a process that in our case was stipulated by our already (toward the end of the project) rich experience in implementing krn with FRANKA Desk. When Melanie demonstrated krn, she had to first get into the cobot now operated through a phase-state machine and train herself to move not only in the right manner, but also to control her arm in a way that she was able to replicate this movement with high precision for three iterations. Thus, attaining a sense for krn now became a process of experiencing the simultaneity of attaining a sense for PANDA and PhaStaProMP as well as disciplining her embodied movements. I suggest thinking of this process in terms of becoming a human-data-assemblage which enabled us to generate the data for transferring a continuous movement onto the cobot.

After the recording and synthetisation of all krn movements, the next step was to use the PhaStaProMP label app and define states. We worked with the three states (1) in, (2) out, and (3) wait. Here, labelling means to categorise movements along these three labels. When labelling, the trained movements are displayed on a virtual model of PANDA, which gave us the opportunity to discard movements which deviated too much from what it should look like. Next, we would run the labelled and therefore trained behaviour on the virtual model. As PhaStaProMP was still in the process of being developed, early in the DRDK project, the movements were too small for the phase state machine to recognise them as movements, resulting from the relation between the friction force necessary to move the cobot's joints, the force needed to actuate the phase state machine, and the force required to demonstrate our knitting movement. Deimel constantly presented us an improved and adjusted version of PhaStaProMP, enabling us to make knitting collaboratively also a test case for his phase-state machine. In turn, we made knitting part of the development of PhaStaProMP, a phase-state machine fitted to the obstacles of realising hand knitting movements on a robot.

After the simulation of the movements, we had to adjust the settings of the execution of krn, like the velocity, but also, and importantly, the controller gains of PANDA regulating the stiffness and softness of each joint in executing the movement. This became central for improving our practice of krn and thus for knitting collaboratively with PANDA. Finally, typing "run on real robot" into the command prompt window, PANDA would start executing the trained

Figure 27: Running the trained behaviour on the virtual robot with blending of phases

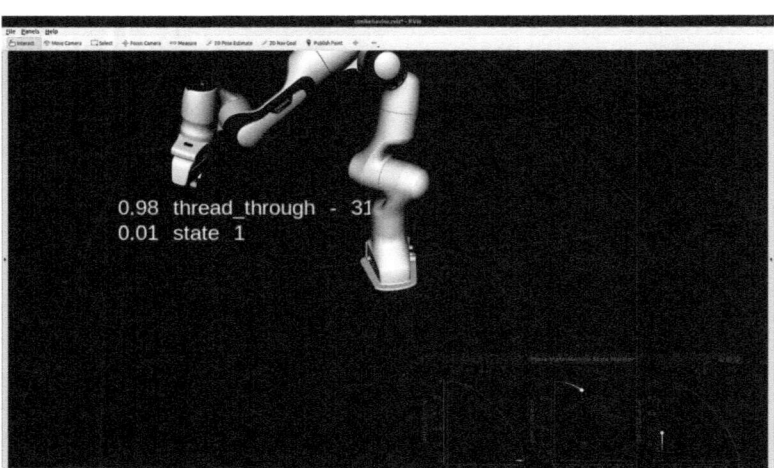

movement. The two most important improvements were: first, that the states would be executed in a blended fashion without fragmentations, and second, that we could adjust the stiffness of the individual joints. After a few tests, we figured out that krn worked best if the fixed end with which the robot was mounted to the table was stiff and the part of the end-effector with which we interacted was adaptable and flexible.

The confinement of the one part to stiffness while liberating another part from stiffness meant to reduce precision and increase adaptability and flexibility. Both the continuous transition of states as well as the differentiated controller gains generated a new quality of collaboration. Knitting with PANDA through PhaStaProMP, Melanie described her experience as less exhausting and more in synchrony with her movements. In contrast, when implemented through the joint motion app, we often encountered the problem that the human knitting with PANDA was either moving too fast or too slow. PhaStaProMP allowed us to develop a common pace, especially based on the difference between iterations (noise programmed by Deimel) and the option to correct the cobot's movement because of the flexible joints. For instance, when PANDA and I knitted collaboratively and lost the yarn in the process of forming a new stitch, I would push the cobot arm slightly to the right in order to adjust to the flow of materials in forming a new stitch, while the cobot would find its way back to its trajectory of krn, operating in a very fluid behaviour. In addition, the continuous instead of fragmented movements also increased this impression. As a result, everyone in the team who knitted with PANDA experienced the circumstances and practice of krn realised with PhaStaProMP as a more authentic form of collaboration than with the joint motion app, potentially moving in the direction of realising hand knitting as a skill together, rather than just executing a purified set of motions. Tentatively arguing, I suggest tweaking the interconnection between an increase of our digital practice and the increase of collaborative capacities, resulting in the improvement of performing collaborative hand knitting as an embodied skill, adhering to the flow of materials. What appears to become tangible here is that the increase of our digital practice in the form of implementing krn through PhaStaProMP not only depicts an embodied skill itself, characterised by the intra-active and rhythmic relating between bodies, data, movements, and materials, but also might open up new avenues for forms of human-cobot co-creation, that is, practicing co-textility.

In this chapter, I did not approach dichotomous relations as given, fixed, and non-disputable common-sense knowledge, but rather as, first, permanently performatively enacted and, second, as itself generative in meaning- and matter-making. Thus, my account of challenging the nature and the power of such dichotomies follows a diffractive methodology, which engages playfully with objects, allowing different aspects to move from the foreground to the background, and vice versa, in order to enable new patterns of relation to emerge. Thus, the three sub-strings are written in such a style of moving aspects of the relation between hand knitting, automation, and digital practice to the fore- and background. From time to time, I as the writer, also felt moved by the strings of knowledge and experience that I am bringing together here—rather than feeling in control and as the only agential entity engaged in this writing. These strings emerged as unasked-for patterns of relating between storylines, forming into string figurings of human-robot entanglements of collaboration through hand knitting.

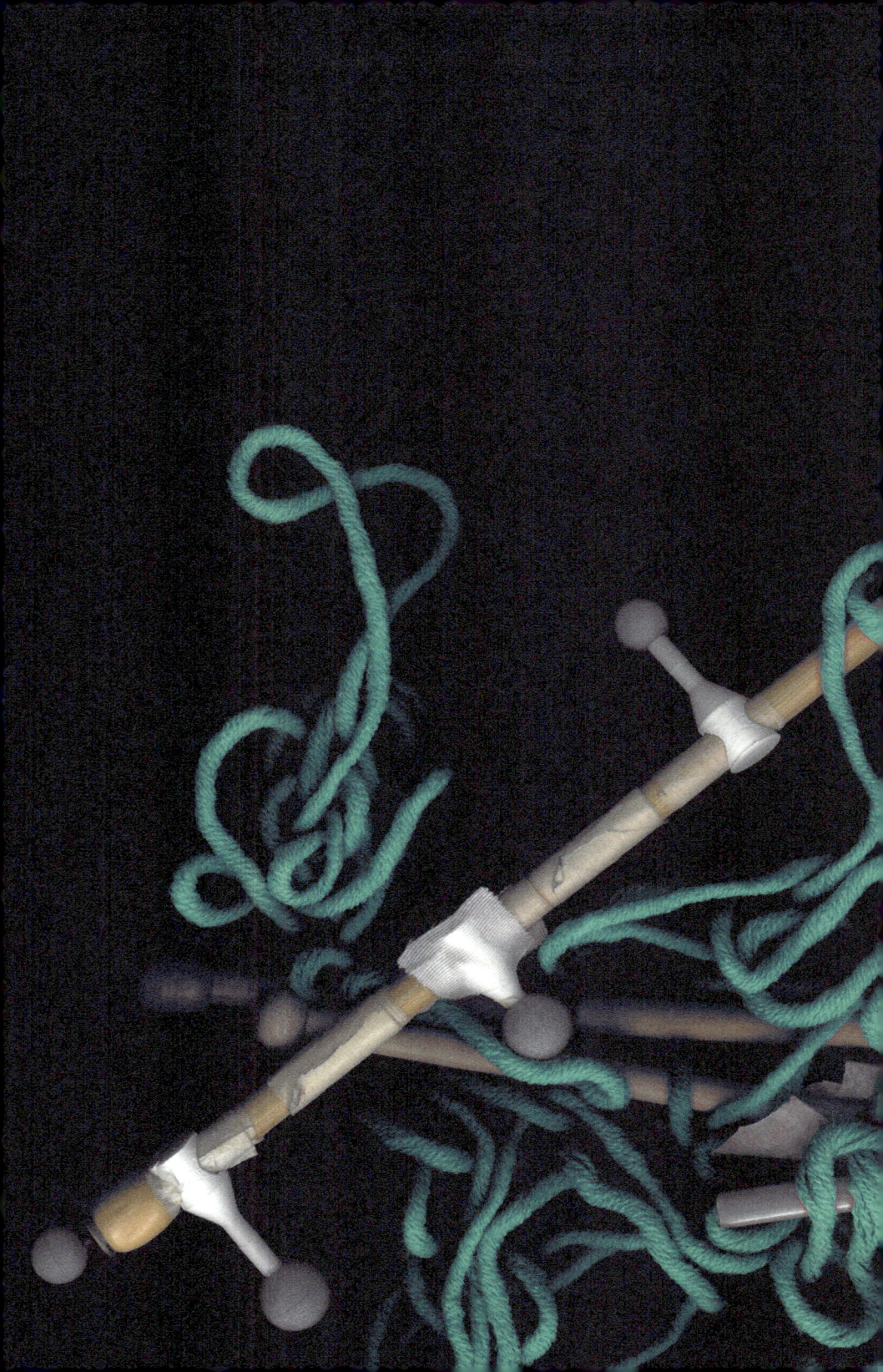

CHAPTER 3
KNITTING TOGETHER: COLLABORATION AS CAREFUL COBOTING

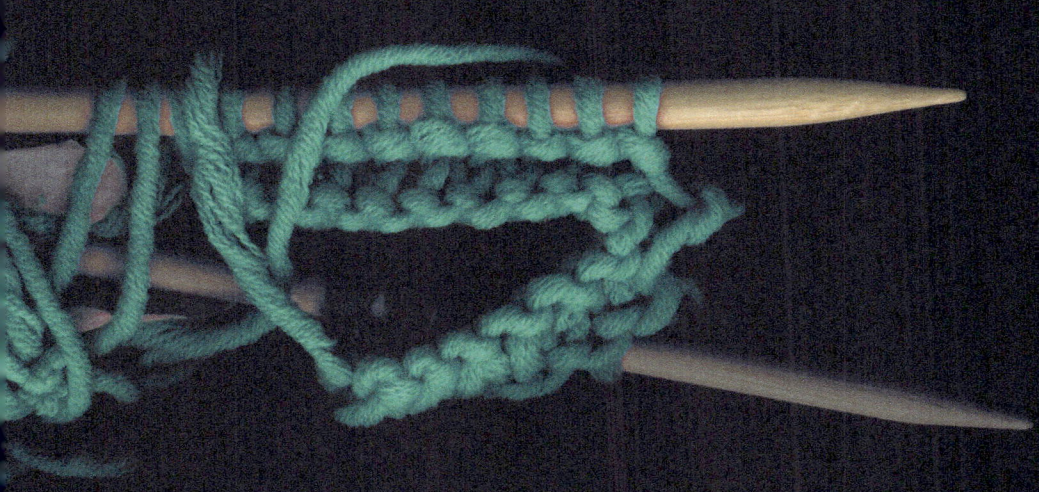

Chapter 3: Knitting Together
Collaboration as Careful Coboting

This third and closing chapter knits together debates on cobots as co-workers or substitute workers (Chapter 1) with the insights from realising knitting collaboratively with a cobot (Chapter 2), in order to generate a knitting-related configuration of present possibilities of collaboration between a cobot and humans, as well as to open up points of departures for speculating about technological futures of re-crafted human-robot-collaboration.

More precisely, I develop an account of human-robot collaboration as *careful coboting*. This term signifies first that HRC is not based on capacities either embodied by the human or the robot, but rather emerges from the entanglement of the cobot and its affiliated humans in their intra-active encounter, captured by the term *coboting*. Second, it defines the analytical sensitivity towards coboting through a stance of *care*, making the situatedness of HRC the matter of care in designing and debating 'our' robotic futures. Finally, careful coboting is a navigational tool for exploring new modes of relating between human and machine through embodied experiences of caring for HRI, as I will continue to argue throughout this chapter.

Robotic knitting diffracts the relations between the cultural technology of knitting and the cutting-edge technology of the robot arm—in each, their multiple meanings—as well as those relations between entities knitting collaboratively with each other. Through this, my approach to human-robot-collaboration enables me to constantly probe ways of relating, and therefore modes of being, through collaborating. Thus, the needles and yarn in the robotic lab became the tools to think and act with: To inquire how 'we' imagine collaboration between humans and robots and to challenge this thinking by probing the enactment of hand knitting as a collaborative task.

Starting with a project that appeared at a first glance either amusing or astonishing, then writing a book documenting my project work, I wove together a critical examination of discourses relevant for understanding the sociomaterial figuration of the robotic (co-)worker with a respecification of material practices through my and my team's engagement in HRC through knitting. Thus, this book shows the myriad and complex dimensions of what started as a somehow funny idea, namely to knit with a cobot, leading to a practice of handicraft as engineering and vice versa. Culminating in the notion of careful coboting as a key concept for re-crafting robotic futures and engaging with robots differently, this notion centrally assembles (1) the neglected labours, (2) the intra-active processes of co-shaping, and (3) the situated, sociomaterial conditions and practices of enactment, including (4) the material and the metaphor of the yarn and textile structures as foundational matters of care for human-cobot-relating.

3.1 String Figuring Storylines & Sociomaterial Configurations

Continuing to be guided by my experiences in the lab, in this chapter I will revisit Chapters 1 and 2 through remembering selected encounters between me, as the PI of the project, and a mostly academic audience interested in DRDK during the time the project was running. Essential to becoming a robotics practitioner through DRDK was to communicate the project in a documenting manner via different media, such as a blog, but also other channels of science communication inside and outside of the Technical University of Berlin (TUB), like the monthly news magazine of the TUB, and Open Science Events, including the open lab events at the MTIengAge lab and conferences of diverse disciplinary audiences. In addition, the project also developed quite a momentum with regard to invitations for project presentations on different occasions and in varying disciplinary circles, ranging from STS meetings with a focus on robotics and AI, to courses in machine learning and textile technologies, to robotic working groups.

These encounters not only shaped my thinking, but they also expanded the frame and re-designed the course of the project. Thus, realising robotic knitting was not confined to the space of the lab, but was rather also effected by the constant exchange through science communication channels. In renarrating selected experiences of presenting and debating the project while we were realising robotic knitting, I take these experiences and debates as my points of departure to outline here the main results of robotic knitting, including the argument for an account of human-robot-collaboration as careful coboting.

I can generalise my experience presenting DRDK, across very different audiences and occasions, as mainly causing two different kinds of reactions, either scepticism or fascination. I experienced these situations as sometimes quite overwhelming. They were not only densely composed of strong emotions of either rejection or enthusiasm, but also had a tendency to make me feel like I was turned into some kind of mediator between the robotic imaginary and a 'robotic reality' for different crowds. Oddly enough, this was prone to make me feel like I was pushed into a position of *objectivity* and 'telling the truth' about robotic presents and futures, while to me, it remained crystal clear that my position has been and remains to be that of a queering witness to string figurations of robotic knitting with the aim to produce situated knowledge on human-cobot-relations in practice *as* theory through knitting collaboratively.

In addition, speaking from my experience of engaging with robotic matters already in my dissertation project on *Robotic Companionship* (Treusch 2015) and since then, what becomes tangible on this and other occasions is that the topic itself, especially through its ubiquity in popular media, is one almost every person has an opinion on. This is not something I find problematic, quite the contrary, it can be an important means for sparking a conversation across diverse groups of persons. However, at the same time, it seems unavoidable for a person living in the Global North to not develop an idea of a robot, what it will look like and how it will behave, how 'we' humans will relate to it and if it is a potential threat or a helpful machine. The challenge I countlessly encountered and encounter is to find a way of avoiding being, again, pushed into a position of *objectivity*, letting my conversations end at the point where I am to tell what current cobots really can and cannot accomplish. Rather, this can only be the point of departure for what I framed earlier as a process of becoming the sand in the gearbox of a too well-oiled *technoliberal* machinery. It is the disruptive momentum of becoming the sand in the gearbox which further allows me to dis- and re-entangle strings of story- and timelines, as well as of sociomaterial configurations of robotic present and futures and concomitant relations of HRI. Instead of delivering clear answers to questions such as "Are the robots really coming?" or "Should 'we humans' reject or embrace this technology?," I offer to become *response-able* (Haraway 1991, Schrader 2010) to the challenge that the contemporary robotic imaginary poses. As Haraway points out, to become response-able is about becoming "answerable [and accountable] for what we learn how to see" (Haraway 1991, 190). Thus, in becoming response-able, that is, becoming responsible through becoming able to respond, to knitting as a craft practice of fabricating worlds and as a practice which entangles matter- and meaning-making, in what follows, I will revisit three different conversational agglomerations which I encountered throughout the project. In this, I draw results from insights presented in the previous chapters, and in navigating through these debates and the results of this book, I will dwell on my technofeminist way between utopian optimism and pessimistic fatalism.

Furthermore, I essentially regard this form of concluding as continuing with playing with strings, following them, mapping nodes and paying attention to (unasked-for) patterns of interference that are constantly moving from the background to the fore and vice-versa. Writing this sentence, I sit in front of a green cardboard box in which we used to store our needles and balls of yarn, as well as knitted artefacts, during the project, where they remain now that DRDK has ended and we have had to move out of the robotic lab. I open this box, grab two balls of yarn and start engaging with individual threads, caught up in the unavoidable mess produced by the unravelled threads in the box, while re-visiting scenarios of the project that form a node and a pattern to hold nodes and patterns of yarn from the project in my hands. Playing with strings of yarn, following this string leads me to another one, one that enabled me to find the words to express the complexity of knitting collaboratively with a robot.

"But—How is this Feminist?"

One articulation of scepticism and sometimes even rejection towards the project, which—to my surprise—I have encountered many times, relates to the feminist dimensions of the project. In my view, it should have been obvious that my project was deeply rooted in technofeminism with its foci on (1) epistemological inquiries of what can be known by whom and what that are always entangled with ontological dimensions of knowledge practices, including agential configurations and sociomaterial conditions of knowing, and (2) on exploring possibilities for making a difference in the design and development of technologies that pivots around the first focus. Further, I regard a diffractive methodology in its aim to find the cracks in the canon from the position of the queering witness as emblematic for this. However, when communicating the project outline and its goals to various audiences, I was facing one re-occurring challenge, namely being asked: "But—how is this feminist?" When I was asked this the first time, I was baffled, tempted to answer with "But—how is this not feminist?" and it took me a moment to understand where this question was coming from. More precisely, I figured out two different expectations towards the project from which these questions were raised: first, the expectation to target the unequal representation and access of women* in the field of robotics and computer science in Germany, and second, the expectation to show that robots are not and will never be ready to become (human-like) cobots, but that claims of a robotic future with co-workers is rather a techno-fantasy emerging from a predominantly White and male robotic culture that 'we feminists' should dismiss—at least more than I do. The assumption around the first was that bringing females with knitting needles into a robotic lab in which exclusively male engineers are working, that I would perpetuate existing sexist stereotypes and cultural codes which associate females with 'soft handicraft' and males with 'hard technology', reproducing the gendered labour division already in place. The bone of contention here was hand knitting and my female-coded team, including me, and I remember conversations with very different persons in which—after convincing my conversation partner that I was indeed a technofeminist—I was continually asked several times: "But—why knitting?" It was not in the least my attempt to convince anyone that either knitting should be regarded as by itself not only a political, but somewhat emancipatory activity, nor that this scepticism towards the project seems to be based on an underlying perpetuation of the very dichotomy between knitting and robotics in its gendered coding. Consequentially, this kind of conversation ended at an impasse.

The second expectation appears to be built on the assumption that realising knitting with a robot means contributing to a techno-solutionism inherent to contemporary technology development, prone not only to attesting that there exists a techno-fix for every societal problem, from de-biasing algorithms, making robots social agents, to data security concerns, but also that contemporary cobot technology also already embodies the potential of cobotic futures. Realising hand knitting then is under the suspicion to be just another contribution to propelling processes of automating every sphere of human everyday lives further. With the project's goal to intervene exactly into this debate, through becoming the sand in the gearbox as a move for performative re-figurations of human-machine collaboration, conversations which reached this point also ended at an impasse.

Writing about these feelings of impasse, I aim at resuming my project work here, while taking into account the affective landscape of these experiences in order to become response-able to and for the two described impasses. The work of Anne Cvetkovich on feelings, handicraft and feminism turns out to be vital in aligning my arguments here further. The cover of her *Depression: A Public Feeling* (2012) is illustrated with Allyson Mitchell's artwork called *Hungry Purse: The Vagina Dentata in Late Capitalism* (see ibid., plates 8-11). Depicted is an opening, assumingly a door frame, completely covered in shag rugs, crocheted fabric and other textile materials, all of them recycled (ibid., 185), making up a structure in different shades of pink mixed with red, black, grey, and white. Looking into the opening, I look first at drapes of fabric, representing "labial folds" (ibid., 186) to then look into a kind of a tunnel which ends in a smaller room, representing the female womb (ibid.). Everything is covered in textile materials, either shag rugs or crocheted and knitted surfaces, for instance structured trough zigzag or arrowhead patterns and, later in the tunnel and room, also granny squares. Close to the opening hangs a huge light pink tassel from the fabric ceiling. The amount of fuzziness of this artwork is so tangible that it can be captured by a photo—the cover feels fluffy and always fascinated and provoked me to imagine how it would feel to touch and walk through the Hungry Purse, taking a break on my way to then end up in the womb, the connected room.

Mitchell is working with materials that might be regarded as *outmoded* (ibid., 185). As Cvetkovich (ibid.) further notes, "for her [Mitchell], the strong and frequently negative feelings attached to objects that are sentimental, cute, garish, cheap, or excessive resemble the feelings associated with both fat girls and feminisms, and this reservoir of shame, abjection, and mixed feelings is a resource for queer reparative strategies." Mitchell thereby engages foremost with negative feelings associated with a certain kind of feminism, namely that of lesbian feminism as a form of feminism which appears to be as outmoded as the rugs, plaids, and other textile materials and the artistic handicraft practices, the crocheting and knitting. Mitchell coins her thinking and art practice as *Deep Lez* (Mitchell cited in Cvetkovich 2012, 186), a perspective of "acknowledg[ing] and address[ing] histories of conflict" (Cvetkovich 2012, 186) and of "seek[ing] to avoid political depression by seeing the past as a potential ally and resource" (ibid., 187). Thus, such acknowledgment does not mean to reinforce these conflicts in their political, collective, and personal dimensions, but rather to open up possibilities of learning from lesbian histories for queer-lesbian presents and futures in a non-dismissive way—a way which avoids shame on a personal level and incapacitation on a collective level. In addition, the affective stance of Mitchell's Hungry Purse literally embodies this queerfeminist reparative strategy: It evokes feelings of comfort and invitedness, that can be experienced as an individual but also across individuals, while the fuzziness of the handcrafted artwork also interweaves feminist pasts, presents, and futures. When Cvetkovich (ibid., 188-189) remembers the happening of a *Public Feeling Event* inside of the Hungry Purse, she describes that "it felt like there was room both to express loneliness and to feel a little less lonely," to then conclude that they have experienced the *utopian performative*.

Mitchell's artwork and Cvetkovich's engagement with it attribute to an account of handicraft and feminism that is different to the craftista movements mentioned in Chapter 2—or at least displays a positioning of its own within the more recent craftista movements. Cvetkovich's work, especially, pivots around a *reparative perspective* (ibid., 10) which works with "legacies of 1970s feminism such as consciousness-raising, personal narrative, and craft" as well as "contemporary queer culture" in its reviving of craft "in order to explore practices of living that both accommodate depression and alleviate it" (ibid., 26). Furthermore, Cvetkovich (ibid., 177) points out that Mitchell's artwork, among others, does not only "embody a reparative response to conflicts within feminism and between art and craft, but the utopian spaces of their large-scale installations produce a reparative experience of depression by literally engaging the senses in a way that makes things feel different." I am intrigued by the experienceability of reparation realised by the overly textile-decked example of the Hungry Purse. Furthermore, revisiting the course of DRDK, I feel like the balls of yarn distributed over the robotic lab might be read as evocative objects in such an affecting sense. Could robotic knitting be considered as an attempt at establishing everyday practices of probing human-machine relations in the robotic lab that hold the potential to amplify utterances of reparative responses to conflicts within and between feminisms, robotics and craft?

The way I assemble strings and allow interferences to emerge, robotic knitting as a practice and as a technofeminist tool embodies the historical relevance of knitting for not only processes of automation, but also processes of cultural ordering along gender, race, and class. At the same time, it neither attempted to reproduce gendered, racial, and classist patterns of cultural coding, nor to prioritise certain feminist goals over others. Rather, robotic knitting is a practice and tool of navigating between uncomfortable processes, like discovering gaps and conflicts made very tangible to me when, for instance, feeling laughed at as a female-read, queerfeminist knitter standing in a robotic lab or when explaining my project to a few, rather sceptical feminists, and the more comfortable engagements with fuzzy yarns and a multitude of human and cobotic knitters, enjoying the individual and collectivizing pleasures of producing a textile, material artefact collaboratively. In this regard, the performative enactment of robotic knitting might also contribute to a utopian performative, as a space in which interdisciplinary research is about more than finding the solutions to previously defined problems and thus moves beyond technorationality. As a navigational tool, it helped us to orient ourselves in our practices of working towards the shared goal we decided to commit to, namely knitting with a cobot, while allowing to constantly question our chosen means of and concomitant assumptions about reaching this goal.

Robotic knitting works towards rec-rafting sociomaterial configurations of HRI in a material and metaphorical manner. Thereby, hand knitting with yarns expresses my deep commitment towards conflicting realities in their desires, understandings, and practices. I do not suggest that everyone should start knitting in a robotics lab, but rather to open robotics culture up for such performative explorations of human-robot-relations, involving equally 'our' bodies as 'our' senses and affects, and deploying diffraction, the game of cat's cradle, as a methodology for taking on a reparative perspective in technofeminist engagements with 'our' technoliberal, techno-driven worlds. Not in the least, this perspective leads me out of the feelings of an impasse emerging from debates over the feminist aspects of DRDK: The disruption produced by my engagement with yarn in the robotic lab is generative, producing patterns of interference, and reparative, allowing conflicting perspectives to co-exist instead of excluding each other.

"This is Deep Collaboration What You are Doing!"

During my project work at the MTI-engage lab over the course of in total about two years, I was able to witness the increasing interest in robotics as a popular topic. For instance, the lab hosted *Open Lab Days* every first Friday of the month that were only announced on the project's homepage, and I could regularly see groups of up to 10 people showing up in order to learn more about current robotic technologies. This could be students or researchers from other departments of the TUB, and from other Berlin-based universities, but also journalists or curious citizens. In addition, the TUB internal media department showed an increased interest in dropping by the lab and producing short video clips or photos for different occasions, but always for science communication purposes, like advertising a lecture on future technologies. On one such appointment, two persons from TUB showed up who had scheduled their appointment with the head of the MTIengAge lab. Part of my new role when becoming a robotics practitioner encompassed a set of activities, including the engineering of collaborative knitting and becoming active in demonstrating the execution of collaborative knitting. This entailed communicating and representing this project in a manner that can be regarded as usual for projects in robotic tech development. Since we were present and working on the DRDK project that day, we decided to also offer to talk about our project and to show them what we were doing. Despite showing an initial interest in the knitting, it nevertheless quickly became clear that this would not become the focus of their shooting. Instead, the persons in the lab, including my team and me, were asked to enact specific scenes with the cobot, mostly scenes of PANDA handing over or grasping an item. The suggested items ranged from a bell pepper to a plastic water bottle to a smaller ball. Contemplating this situation, I figured out what this scenario delivered that the knitting with a cobot did not: It was presumably immediately intelligible for most kinds of audiences that the robot is giving an item or taking it from a human, and in consequence, that it can grasp and interact in a meaningful way. In addition, 'we', the informed public, are already familiar with these kinds of images of robots and humans. Knitting is not legible in the same manner—as visible for instance in *Figures* 14 and 18. Our knitting practice, first of all, involved not only the cobot and a person, but rather more than one person. Furthermore, the knitting movement (unlike handing over or grasping) consists of a set of complex, smaller sub-movements, and it appears impossible to produce one picture which is able to tell the story of this practice in all its dimensions, that is, a legible output. In this regard, the challenge of depicting knitting collaboratively with a cobot makes tangible what a queer use of the cobot and of the knitting needles and yarn causes: the necessity to ponder about the way 'we' are used to thinking of robots as useful machines, by allowing new images to evolve.

At a first glance, the small knitting movements might appear as rather non-spectacular, especially in the context of HRI and HCI. Basically, when filming or taking a picture, one does not see that much, besides a person next to a cobot and knitting needles with yarn. When taking close-up pictures of the needles and yarn, the bodily arrangements become invisible. Sadly, at this exemplary occasion of taking photos for a science communication purpose, I was not able to intervene in what appeared to me to be a reification of the already well-known images of how humans and robots can relate, including the idea of what a robot is. At this point in time, I also realised that I myself did not really have the words at my disposal to express how and why when looking at it more closely, knitting collaboratively between human and robot turns out to be at least equally spectacular as the handing-over-an-item scenario. My struggle for words and with becoming able to express why it might be worth looking into our robotic knitting, insisting that the complexity and spectacularity of this practice of enacting HCI will become intelligible, made me feel, again, at an impasse.

Re-narrating in Chapter 2.1 the historical relevance of the yarn for digital technologies and automation technologies, and therefore also robotic technologies, it becomes crystal clear to me that the yarn not only matters on a metaphorical level, but also equally on a material level, in re-crafting human-robotic-futures. The yarn cannot be reduced to either dimension, but rather becomes a powerful tool when acknowledging both in their interrelatedness. Thus, depicting the robot with humans and yarn can be said to have a historical legacy of playing a role in the development of automation technologies as the predecessor for digital computer technologies.

Highlighting the historical prevalence of needlework for automation, including digital technologies, however, is not only about recuperating a relation of kin rather than kind between handicraft and high-tech, viewed as dichotomous spheres. Rather, it is also about exploring the interconnectedness of practices at the textile/technological interface. What becomes very vivid in Chapter 2.2 is that hand knitting displays a challenge for HCI. This largely results from the gap between hegemonic ideas of automation and the everyday practices of realising automation in a robotic system. Entering the lab as an FSTS scholar specializing in human-robot relations, I was well aware of the circumstances of HRI, how messy and ridden with experiences of machine failure it can be to operate a robotic system and to realise the execution of a certain task with the robot. Practices of engineering a robot system to execute a certain behaviour rely on the alignment of things, persons, affects, and actions (Suchman 2007, Alač 2009, Treusch 2015, Lipp 2019).

Thus, the challenging character of hand knitting as a task also works with the potential to tweak the picture of a frictionless operation of a robot, as well as that of engineering as the frictionless mastering of hard- and software. It is not only about producing images of a robot doing different things than what 'we' are used to as citizens informed by the idea that "the robots are coming"—a narrative which evokes the robot as essentially human-like co-worker or substitute worker. Rather, it is equally about making a difference in the doings of HRC in particular and HRI in general. Hence, depicting robotic knitting is about picturing the diffraction of frictionless automation through neglected practices of engineering, involving care on an emotional and corporeal level as it shows, for instance, in practices of attaining a sense for krn.

Again, however, one challenge remained how to communicate the essential aspects of picturing robotic knitting in a more accessible manner. During a research stay in Copenhagen, I was invited by Kasper Støy to present robotic knitting at the REAL (*Robotics, Evolution and Art Lab*) at IT University. This was a very pleasant experience as the context of REAL, first, is that of an interdisciplinary robotic lab, and therefore open to the idea of an FSTS scholar becoming a robotics practitioner, and second, as I had already been in conversation with the roboticist Kasper who was not only interested in this specific project work, but is equally interested in bringing humanities, especially FSTS, in conversation with robotics. After my presentation, Kasper suggested understanding knitting collaboratively with a robot in terms of *Deep Collaboration* (*DC*).

I was immediately intrigued by this framing. It depicts a twist of the hegemonic AI machine learning paradigm, namely *Deep Learning* (*DL*), through a very hands-on scenario, namely the knitting hand and knitting gripper. DL, basically, is grounded in the machine learning technique of so-called (*artificial*) *neural network*s. Such artificial neural networks are conceived of as modelled after "the mechanism of learning in biological organisms" (Aggarwal 2018, 1). This "biological mechanism" encompasses, first and foremost, the neurons and synaptic connections (ibid.). In short, applying this model in machine learning means working with "computational units [that] are connected to one another through weights" (ibid.). This assemblage of artificial neurons and artificial synaptic connections leads to an artificial neural network which, in short, consists of the networked input neurons, an unknown middle layer of abstract operations, and the output neurons. For a neural network to become a functional method of data processing, it has to be trained with so-called *training data* as a foundation for the model generalisation, that is, the capacity to generalise models from the training data. After that, the DL technique is expected to produce the most highly reliable output data when properly trained, and to be especially suitable in operating with large data sets, so-called *big data* (ibid., 4). Despite this very technical description of DL, Zweig (2018, 33, translation: PT) reminds us that "systems of algorithmic decision making should...not be considered in isolation, but always as part of a sociotechnical overall structure"—a fact that can be easily neglected in the selection, training, and modelling of DL techniques.

Even though Deep Collaboration points towards the contemporary figuration of the robotic imaginary and the larger framework of AI with its focus on Deep Learning and its concomitant modes of datafication, formalisation, and statistical (machine) learning, it, at the same time, distances itself from this very framework and brings to the fore the *collaborative as physical, embodied and social-cultural* nature of AI, and especially of interactive machines such as the cobot. Furthermore, it does so on a very hands-on level, namely through the collaborative handling of yarn and needles as a technique and the forming of new stitches as a skill. Thus, I regard robotic knitting as foundational for DC and the latter as a necessary completion of DL.

As illustrated in Chapter 2.2, the neglected skills of hand knitting encompass adherence to the process of forming new stitches, including engaging with the flows and transformations of the materials of hand knitting. Resulting from this is an account of hand knitting as exemplary for being possessed by action—a form of emotional and bodily care for the flows and transformations of materials as well as the activities of all agents and things aligned in this process. This quality of robotic knitting articulates itself in working with yarn as a tool for making the practices and practicalities of collaborative knitting a matter of care. Hand knitting and collaborating with PANDA both rely on tactile and tacit knowledge, both are precarious in nature and require a person to be creative, tinker, and test improvements spontaneously. Hence, diffracting engineering and handicraft, their relations have to be reconsidered as that of kin instead of kind. DC captures precisely these dimensions of robotic knitting.

Further, I imagine Deep Collaboration to establish into a paradigm not only of re-crafting engineering, but equally of working towards new images of HRC, making the spectacular nature of the micro-practices of collaboration tangible to different audiences. In the case of robotic knitting, however, this always encompasses acknowledgment of the potential of hand knitting for developing and practicing DC. Thus, my account of DC also necessarily involves challenging contemporary ideas of knitting between knitting as an outmoded, and "most boring cultural technology of the world" (Wallnöfer 2011,

47, translation: PT) and knitting as means for *technoliberal* subjectivity, propelling an *aesthetic capitalism*. The *textility* of knitting reveals the potential to amplify the utterances of a reparative response-ability to contemporary conflicts between digital and analogue, robotic automation and human (creative) work, as well as *technorational* solutionism and a diffractive inquiry across disciplinary boundaries along the process-oriented practice of attaining a bodily sense of robotic knitting, and following flows and transformations of yarn, hands, and grippers, all working together.

"You Invented the Embodied Turing Test!"

Even though I was immediately intrigued by the idea of having a robot arm knit, the realisation of this idea also made me curious about how the complex task of knitting could be automated through a robot arm, and how this might change the activity of knitting, but equally my concept of automation. Would this still be the handicraft of 'human knitting' or would it become something else?

Already the very first practice of knitting in the robotic lab propelled me to develop robotic knitting as a technofeminist tool for not only re-engineering collaboration with robot technologies in a hands-on manner, but also for re-crafting contemporary robotic cultures of striving for robots as co-workers. Notably, it was the initial attempt at turning knitting into data, as well as the concomitant enactment of this data, which lead me to contemplate hand knitting with a cobot through different eyes, making me wonder how robotic knitting is more than a simple transference of movements onto a robot, but rather the beginning of an exploration of practices of knowing and being in the cobotic lab. In this view, the seemingly useless task of knitting with a robot can become the utopian performative.

In contrast to this experience, machine automation is held as excluding human creativity by necessity. Thus, a core challenge then became to re-join both practices in a non-exclusive manner. However, this challenge articulated in different forms. This became especially tangible to me through one situation. Quite early in the project, I decided to participate in the yearly *Open Science Event*, the so-called *Lange Nacht der Wissenschaften (LNDW)*, at TUB in the following year. Part of this was to submit a short description of what I would like to offer (a robotic knitting station) and an accompanying picture. At that time, I only started to think about images of robotic knitting and was quite fascinated by a couple of pictures that Katrin had just shot. One of them showed two PANDAs facing each other and the camera, while the robot on the left holds a knitting needle with a red knitted piece on it and the robot on the right holds two long metal knitting needles in its gripper that are plunged into the red ball of yarn. The background is bright green—a background paper that was already available in the lab. I thought this was a nice arrangement of robots with needles and yarn—what I did not think was that this picture would—from then on and until this book—determine and define the appearance of DRDK.

Later on, I noticed that the picture was liked by the TUB press department, and from there even evolved into the guiding theme for the LNDW in 2019. The latter meant that it was on the cover of the LNDW magazine of the TUB, the different sections of the magazine were separated by various shots of the described scenario, posters announcing the LNDW with one or two robot arms with needles and yarn could be found all over Berlin, and finally, even the president of TUB, Christian Thompson, drew on the *knitting robots* in his foreword to the LNDW magazine, speaking of the knitting robots as providing a red thread through the programme of the LNDW. I was at the same time amazed and puzzled. It was amazing how much the university seemed to identify with a quirky, technofeminist project; but at the same time, I was also puzzled as the situatedness and outlines of DRDK did not become visible, as I quickly realised—neither through the image nor the texts around it. To me, this was a very interesting experience in science communication. Again, one part of the project, namely that it has something to do with robots, is made visible while making invisible the other part of the project, namely that it is a technofeminist intervention. In my view, this is not about silencing the technofeminist part, but more about the customs in science communication which, as already explored in the subsection above, seem to entail encountering the gaze of the hegemonic robotic imaginary. What the picture (which I delivered in the first place) transported in result, was the image of a robot as emblematic for the potential to possibly automate every sphere—and therefore also every task—of human existence. It became the picture of a story on multi-capable robots that can even knit. Here, I found myself again in the position of taking on the role of the witness to robotics—however, in this case, not to tell how far actual robots are from the narrative "the robots are coming", but rather to affirm how far robotic tech development has come when the robots can even knit. Clearly, I found myself again at an impasse.

Facing this dilemma as part of a more foundational mismatch in aligning robots as cultural figures, contemporary robotics, and the hegemonic robotic imaginary, my task was again to become response-able to conflicts as a technofeminist robotics practitioner. This involved, first, producing flyers for the actual event of the LNDW, and second, being present with my team, Melanie, Anne, and PANDA throughout the LNDW. The flyer re-situated the project and aimed at making visible what did get lost in the image of the two robots with yarn and needles, while over the course of six hours, Melanie and Anne were knitting collaboratively and I was speaking into a microphone, explaining the practice of robotic knitting. As the robot that can knit was the guiding theme of the whole event, we were granted a prominent spot in the main building of the TUB. After the first hour, we decided to run hourly demonstrations. Groups of up to 80 persons came to these demonstrations. Thus, the interest in the project was overwhelming. Simultaneously, I considered this also a great chance to engage in conversations with the persons visiting our demonstrations in order to debate with them the outlines of the project. Further, I hoped to gain insights on what was needed on a visual and textual level to make the project work tangibly for the persons I spoke with—of course, to examine the intelligibility of robotic images would be a project on its own (see Hasse 2019), but, nevertheless, I was curious to hear about the expectations that our visitors had.

In the end, a very large number of persons clearly had the expectation that our demonstrations would show two robots knitting with each other and expressed their disappointment. However, there were also others. What especially caught my attention were grandmother-grandson tandems. It happened a couple of times that a grandmother came up to me, explaining that she came to this demonstration because her grandson is fascinated by robots, while expressing her fascination with needlework. These situations proved to be of special value for communicating the project work. I could easily reduce this situation to one of a gender-stereotypical expression of interest, but through a reparative lens, I discovered that these situations brought the potential of an intergenerational conversation about both needlework and robots that then allowed me and my team to argue why and how both are interconnected and to explain what DRDK does. In general, I was surprised how many persons—even those disappointed that we did not show two robots knitting with each other—were keen on learning more about our exploration of HRC through knitting. Thus, in the end, the impasse turned into a surprising experience of not only communicating the project work, but also of generating a more complex understanding of my role as a technofeminist robotics practitioner, engaging in the usual formats of science communication. Nevertheless, the two robots potentially knitting together, representing DRDK, continued to bother me. How could I re-integrate the collaborative aspects into that image?

During a presentation of DRDK at the *GeDIS* (*Gender/Diversity in Informatics Systems*), at Kassel University, led by Claude Draude, Claude suggested that the implementation of robotic knitting in which the robot is supposed to learn 'human hand knitting', that is, how to knit like 'us' humans, depicts an *embodied Turing Test*. In what follows, I will develop this idea further in order to explore the possibilities to tweak the idea of automation as substituting humans towards a more collaborative understanding. Developing impulses for a re-crafting of robotics, I suggest a re-working of the popular and well-established *Turing Test* through robotic knitting, inspired by Claude. Based on Turing's challenge of 'our' humanist scientific foundations in conceptualising the machine Other, the Turing Test has advanced into the gold standard of determining the human-like intelligence, and thus successfulness, of AI. However, this test and how it is implemented focuses on purely cognitive, immaterial terms. Hence, the material and embodied nature of collaboration through hand knitting appears as an ideal example to revisit the Turing Test in terms of embodiment and embodied affects.

Again, as a hand knitter, I can say that yarn is quite a stubborn material when having to master it with needles. It can start dissolving itself into individual threads, it can form knots when unravelling, and it can be either too slippery or too cumbersome when forming new stitches. Hand knitting, hence, requires a precise handling of both the yarn and the needles. This requirement makes it a human-exclusive activity—in contrast to its automated form, performed by knitting machines. In fact, it was by no means necessary for me to invite a cobot to join me in mastering yarn and needles successfully. Thus, I assumed from the very beginning of DRDK that robotic knitting as a task requires a high degree of adaptability between human and robot—a requirement which also shows very vividly in my recounting of realising robotic knitting in Chapter 2.2.

Beyond this, I suggest contemplating about this adaptability as a possible way out of defining human-likeness as the only possible form of reaching a mutual understanding between human and robot. Thus, I am curious about the idea of connecting human knitting performed by a robot, in its potential to conflate human and machine action, to the Turing Test. The latter, in short, was proposed by Alan Turing in 1950 and tests the indistinguishability between human and machine interlocutors. Notably, and as Draude (2017,

191) reminds me, "before Turing develops a scenario for human-machine interaction, he invents a gender imitation game, in which different roles are attributed to each gender." The *Imitation Game* first invented a situation, in which one person (the interrogator) engages in a conversation with two others (a man and a woman) via typewritten questions and answers, while located in separated rooms, and then must determine the gender of each person. Furthermore, the goal was to confuse the interrogator as the two others were instructed to both take on the female-coded role of assisting the interrogator and therefore "both players try to convince the interrogator that they are the woman" (ibid.). For Turing, if a man can convinced the interrogator that he is a woman, then "the imitation of the woman by the man may be replaced by the imitation of the woman by the machine" (ibid.). Foundational to both the Imitation Game and the Turing Test are to separate between sign and body/materiality as the basic principle to enable such imitation. As Draude (ibid.) resumes, "according to the Turing Test, the sign is treated ... as freed from the connotations, restraints, and limits that an embodied existence brings along." In this regard, Turing's work can be read as potentially encompassing a queering of boundaries between meaning and matter, female and male, and human and machine. Elizabeth A. Wilson (2010) argues in a similar vein when she attests that when Turing raised the question of *Can machines think?* (Turing 1950, 433), he opened up the possibility of challenging modern thought with its concepts of thinking and intelligence as solely properties of the human subject (ibid., ix). Against the backdrop of these readings, how could the practice of knitting with a robot contribute to the queering and curious nature of the Turing Test?

First of all, it is pivotal for robotic knitting to take up the queer momentum implied in the Turing Test as a mostly neglected dimension. It basically works towards a de-essentialisation of a determinist relation between matter and meaning. At the same time, the momentum of curiosity equally deployed by the Turing Test, opens up the realm of the unthinkable. Together, my account of the Test propels the probing of human-robot relations beyond the category of the human-like as that which allows a mutual understanding between humans and robots. With this, I bring to the fore the Test's capacity to de-couple form and function and to allow the unthinkable to take shape while I insist on the embodied nature of experiencing human-robot relations *differently*.

Robotic knitting moves beyond narrative formations which operate with universalising claims about who will be affected and how by an increase of automation technologies. It does so by challenging the very notion of automation. Passing the embodied Turing Test is a prerequisite for this. It allows me to develop a more capacious understanding of *collaboration*, and thus also of the relation between the cobot and the different humans involved, in terms of *careful coboting*: Taking into account the practices and practicalities of collaborative knitting means to acknowledge the sociomaterial configuration of robotic knitting. It opens up a field between *techne* and *logos* through which the transference of a skill onto a robot stipulates a re-signification of labours and affects invested into HRC, that works more in line with an intra-active than interactive paradigm. The latter is geared at assembling the actors and actions in their entangled nature in order to foreground the co-shaping character of human-machine relations, from which entities with boundaries emerge. The question of what a robot—and in this relation, also a human—is, then, cannot be cut off from the multi-faceted enactment of relations as well as their experienced reality *in situ*. This also opens up possibilities to re-pose questions of responsibility for the distribution of labours between humans and robots: From gestures of propelling either promises or fears to becoming accountable for and through situated practices of relating.

Not in the least, these insights are diametrically opposed to discourses which confine 'the human' to the *technoliberal* subject in danger of a White loss, as described in Chapter 1. Careful coboting places human and robot together, underlining their dependent and entangled relation—neither 'the human' nor the robot can be erased from this picture. Rather, what needs to be erased in order to engage in careful coboting is the very logic of technoliberalism, including the pattern of surrogacy in its oppressive operations of ordering. Robotic knitting is constitutive of a utopian performance that articulates in the mundane practices of being possessed by hand knitting across human and robot, possibly evoking human-machine co-creativity and therefore also re-joining human creativity and machine automation. It cannot provide one picture that captures an ontological essence of HRC, but rather has to stay with the trouble of constantly engaging with mundane practices of HRC, or, put differently, with the realisations of Deep Collaboration and with passing the embodied Turing Test.

3.2 Careful Coboting through Hand Knitting – and Beyond

> ... the thing has the character not of an externally bounded entity, set over and against the world, but of a knot whose constituent threads, far from being contained within it, trail beyond, only to become caught with other threads in other knots.
> Ingold, *Bringing things to life*, 4

Bringing yarn to the robotics lab is more than what appears at first sight as an amusing and playful, but nevertheless interventionist, endeavour. It is also not limited to challenging contemporary automation technologies through the implementation of an unconventional task of HRC. It centrally aims at re-crafting the human-robot relations of collaboration—through first taking into account that the human-robot interface is loaded with cultural meaning, and therefore not only reproduces existing power relations, but also determines how robots and humans can relate in imaginations of HRI, but also *in praxis*. Second, and based on that, such a re-crafting includes to reclaim care in its emotional labours and corporeal forms as a substantial dimension of collaboration between humans and robots, while establishing that this care is not based on well-known images like the infant-caregiver metaphor, but rather understood as a foundational practice of engineering human-robot-relations. The latter is where practices of hand knitting and practices of engineering in their textility of creation, including their precarious, provisional and tactile nature, conflate.

The category of the human-like should supposedly evoke visions of specific figures, thought of as enabling socially meaningful relations with robots, such as the maid or that of the slave. At the same time, the ways in which these figures are charged with social relations of power, exclusion, violence, and oppression, but also of resistance and overthrowing, are largely ignored. However, acknowledging existing power relations in the development of human-likeness is not sufficient in overcoming the restrictions of the category of the human-like. Rather, robotic knitting, in its disruptive and generative engagement in forming new stitches, not only explores the sociomaterial limits of that category, but also traces the potential to transgress these limits by advancing structures of relations that are not yet imaginable. In this sense, the here carefully assembled, detailed account of knitting with a robot—as a practice of coboting—illustrates the ways in which the technological realisation of abilities in the robot cannot be exclusively confined by the category of the human-like. Further, essential for transgressing the limits of the category of the human-like is also the acknowledgment of the ways in which HRI cannot be planned or captured through a fixed set of affordances and constraints embodied by a concrete artefact. Rather, the cobot's capacities exceed the pre-planned 'pure functionality' of a device. They emerge from the intra-active encounter between robot and human. Illustrating this intra-action in Chapter 2.2, then, shifts the focus from the human-like as the primordial category of human-legibility to the multi-dimensional practice of enacting collaboration with a focus on care as a stance of caring for how people and things matter together. While these assembled factors of collaboration are foundational for becoming socially meaningful, they are mostly neglected. Assembling the neglected factors in turn is then a practice of care.

Robotic knitting works with and through the complex interrelations of culturally charged attributions, agencies, affects, and the embodied practice, as well as the experience of human-robot relations of collaboration, by moving beyond the usual scenarios of the helpful or useful robot. Essential to this is to remember the historical traditions of needlework as a means of patternmaking and communication that are foundational for 'our' contemporary everyday technologies (see Monteiro 2017), as well as the gendered coding of handicraft in opposition to technology development. Furthermore, even though realising collaborative knitting between humans and a cobot might appear at first sight as a clear-cut goal, central to this project was the constant examination of the everyday practices of engineering through which we, the interdisciplinary team, were implementing this goal, as well as becoming attentive to the host of labours, materials, affects, and agents involved in this process. Robotic knitting thus serves as a tool for probing taken-for-granted knowledges, and practices of engineering such a goal, while at the same time, it also functions as a tool for re-engineering and telling a different story. It enabled us to performatively test the handling of yarn and knitting needles by humans and a cobot in order to not only probe, but also re-craft human-robot collaboration. This encompassed translating the negotiations of what HCI means and could mean into experienced reality of human-machine co-creatvity. Hence, robotic knitting is about creating an account of HRC that blurs the boundaries between subject and object, the productive and unproductive, and value and valuelessness, and thereby is geared at advancing structures of relation that were unimaginable before. The conversations that ended at an impasse, which I recounted in this closing chapter, bear witness to the fact that articulating human-cobot relations of careful coboting through knitting is in need of the foundational work on advancing structures of relation that were unimaginable before the project.

Robotic knitting turns out to be (1) the complex effort of re-joining legacies, (2) allowing counter-intuitive relations of kin instead of kind between seemingly dichotomous realms to unfold through a queering of use, (3) a momentum for a yarn-related process-ontological account of generating a more capacious vision and version of collaboration between human and robot, and (4) opening up a field of tensions between the cobot and knitting, in which the cobot is simultaneously a technology rooted in and emerging from normative orderings of present-day capitalism, while hand knitting evokes the potential to subvert this very normative order.

The partial nature of the knowledge produced here does not pertain to a relativist or fatalist stance—rather, quite the opposite: Situating knowledge claims allowed me to cut through what I frame as the ubiquitous and confusing sociotechnical formation of the contemporary robotic imaginary. This then builds the backbone of the book: To separate strings of story- and timelines, as well as of sociomaterial configurations, in order to diffract these strings on entangled narrative as technological as narrative levels. Realising robotic knitting then is reached through a set of practices of diffracting the different strings of robotic knitting, conceived of as human and more-than-human generative nodes, accentuating the formative, co-shaping power of materialising locations in a net of actors. This perspective became vital for me to establish a practice of care in the dis- and re-entanglement of threads. I frame the set of technofeminist practices around robotic knitting in terms of a careful coboting, which deploys (1) a perspective of reparative response-ability, (2) the paradigm of Deep Collaboration, and (3) the embodied Turing Test.

The wool and the knitting needles served me as literal and metaphorical tools to bring the well-oiled machinery of technoscientific envisioning, in its determinism and implicitness, to a halt. This halt then allowed me to explore different possibilities for HRC, not only in theory but also in practice, entangling both theory and praxis in an experimental manner. This exploration is based on the belief in the openness of the robotic future as well as the need for making a difference in 'our' robotic futures that is not about finding more adequate representations of what visions and realisations of collaboration between humans and robots really looks like, but to radically open up possibilities for new realities of human-cobot-relating—possibilities that might not have been explored before and that cannot be determined prior to probing them.

Laying out the strings of robotic knitting here, I wish for them to get caught up with other threads and form new nodes, constantly queering the what and who of a proper use and working towards a more tangible and situated debate about human-robot futures also beyond robotic knitting. Pivotal for this is to continue insisting on the circumstance that 'our' human-robot futures have not yet been written, but rather are indeed open. Human-robot relating is a practice of culture, working with operations of hierarchisation and in- and exclusion. However, it not only matters how and under which auspices 'we' tell the story of a dawning of an inevitable robotic future, but it is also, and importantly, the enactment of the very mundane relations through which robots are woven into 'our human' sociomaterial fabric that matters. Thus, making a difference in relating is about engaging in careful coboting.

Finally, the last two pages of the book depict the collection of knitted artefacts that were produced during the project. As every knitted artefact embodies the history of its production, I regard the knitted artefacts of DRDK as a manifestation not only of its story, but also of the story of careful coboting. We did not produce nicely uniform knitted artefacts. Instead, they are marked by holes which I understand less as failures, but rather as exceeding such simple assumptions. They are *Leerstellen* (gaps), literally translating to empty spaces. Becoming accountable for what and how I see, I speculate about these Leerstellen as signifying the openness of what robots and how they can be woven into 'our' sociomaterial fabric, and who should participate, as is articulated through DRDK and beyond. These Leerstellen interweave technofeminist robotic pasts, presents, and futures as a matter of careful coboting.

Bibliography

Adam, Alison 1998. Artificial Knowing: Gender and the Thinking Machine. New York: Routledge.
Aggarwal 2018. Neural Networks and Deep Learning. A Textbook. Wiesbaden: Springer.
Ahmed, Sara 2004. The Cultural Politics of Emotion. New York: Routledge.
—. 2019. What's the Use? On the Uses of Use. Durham: Duke University Press.
Alač, Morana 2009. Moving Android. On Social Robots and Body-in-Interaction. Social Studies of Science, 39/4, pages 491 – 528.
Arantes, Lydia Maria 2017. Verstrickungen. Kulturanthropologische Perspektiven auf Stricken und Handarbeit. Berlin: Panama.
Atanasoski, Neda and Kalindi Vora 2019. Surrogate Humanity. Race, Robots and the Politics of Technological Futures. Durham: Duke University Press.
Barad, Karen 2007. Meeting the Universe Halfway. Durham: Duke University Press.
Barla, Josef 2019. The Techno-Apparatus of Bodily Production: A New Materialist Theory of Technology and the Body. Bielefeld: transcript.
de la Bellacasa, Maria Puig 2017. Matters of Care: Speculative Ethics in More than Human Worlds. Minnesota: University of Minnesota Press.
Birhane, Abeba and Jelle van Dijk 2020. Robot Rights? Let's Talk about Human Welfare Instead. 2020 AAAI/ACM Conference on AI, Ethics, and Society (AIES'20), February 7–8, 2020, New York, 7 pages. https://doi.org/10.1145/3375627.3375855
von Bose, Käthe and Pat Treusch 2013. Von 'helfenden Händen' in Robotik und Krankenhaus. Zur Bedeutung einzelner Handgriffe in aktuellen Aushandlungen um Pflege. Feministische Studien, 2/2013, pages 253 –266.
—. 2018. Keime, Zeitdruck und Roboter als Helfer für alle: Interferenzen zwischen materiell-diskursiven Fürsorgepraktiken in Krankenhaus und

Robotiklabor. B. Binder, C. Bischoff, C. Endter, S. Hess, S. Kienitz (eds.): Care. Praktiken und Politiken der. Opladen: Budrich, pages 192—208.

Bratich, Jack Z. and Heidi M. Brush 2011. Fabricating Activism: Craft-Work, Popular Culture, Gender. Utopian Studies, 22/2, pages 233—260. doi: 10.5325/utopianstudies.22.2.0233

Brooks, Rodney 1991. Intelligence Without Reason, Artificial Intelligence Laboratory. A.I. Memo, No. 1293. http://people.csail.mit.edu/brooks/papers/AIM- 1293.pdf

Butler, Judith 1990. Gender Trouble. New York: Routledge.

Cockburn, Cynthia and Susan Ormrod 1993. Gender & Technology in the Making. London, Thousand Oaks, New Delhi: Sage.

Cvetkovich, Anne 2012. Depression. A Public Feeling. Durham: Duke University Press.

Deimel, Raphael 2019a. Reactive Interaction Through Body Motion and the Phase-State-Machine. EEE/RSJ International Conference on Intelligent Robots and Systems (IROS), Macau, China, 2019, pages 6383—6390. doi: 10.1109/IROS40897.2019.8968557

—. 2019b. A Dynamical System for Governing Continuous, Sequential and Reactive Behaviors. Proceedings of the ARW & OAGM Workshop 2019, 6 pages. doi: 10.3217/978-3-85125-663-5-15

Devendorf, Laura and Daniela K. Rosner 2015. Reimagining Digital Fabrication as Performance Art. CHI EA '15: Proceedings of the 33rd Annual ACM Conference Extended Abstracts on Human Factors in Computing Systems, April 2015, pages 555—566. doi: 10.1145/2702613.2732507

Dourish, Paul 2001. Where the Action Is: The Foundations of Embodied Interaction. Cambridge: MIT Press.

Dourish, Paul and Genevieve Bell 2011. Divining a Digital Future. Mess and Mythology in Ubiquitous Computing. Cambridge: MIT Press.

Draude, Claude 2017. Computing Bodies: Gender Codes and Anthropomorphic Design at the Human-Computer Interface. Wiesbaden: Springer.

Draude, Claude, Goda Klumbyte, Phillip Lücking and Pat Treusch 2019. Situated Algorithms: A Sociotechnical Systemic Approach to Bias. Online Information Review, 44/2, pages 325—342. https://doi.org/10.1108/OIR-10-2018-0332

Drolshagen, Ebba D. 2017. Zwei rechts, zwei links: Geschichten vom Stricken. Berlin: suhrkamp.

Frankjaer, Raune and Peter Dalsgaard 2018. Understanding Craft-Based Inquiry in HCI. DIS '18: Proceedings of the 2018 Designing Interactive Systems Conference, June 2018, pages 473—484.

Gemeinboeck, Petra and Rob Saunders 2016. Creative Machine Performance: Computational Creativity and Robotic Art. Proceedings of the Fourth International Conference on Computational Creativity 2013, pages 215—19. https://computationalcreativity.net/home/

Graf, Philipp and Pat Treusch 2019. „Und dann fängt er einfach an, den Ball auf den Boden zu werfen!" Einsichten in Interaktionsverhältnisse als Statusverhandlungen mit dem humanoiden Roboter Pepper. J. Hergesell, A. Maibaum and M. Meister (eds.): Genese und Folgen der „Pflegerobotik". Weinheim and Basel: Beltz Juventa, pages 46—61.

Gross, Shad, Jeffrey Bardzell and Shaowen Bardzell 2013. Structures, Forms, and Stuff: the Materiality and Medium of Interaction. Pers Ubiquit Comput, pages 1—13. doi: 10.1007/s00779-013-0689-4

Gunkel, David J. 2017. Mind the Gap: Responsible Robotics and the Problem of Responsibility. Ethics Inf Technol, 14 pages. doi: 10.1007/s10676-017-9428-2

—. 2018. Robot Rights. Cambridge: MIT Press

Hakli, Raul and Johanna Seibt 2017. Sociality and Normativity for Robots. Wiesbaden: Springer.

Haraway, Donna J. 1985. A Manifesto for Cyborgs: Science, Technology, and Socialist Feminism in the 1980s. Socialist Review, 80, pages 65—107.

—. 1987. Donna Haraway Reads 'The National Geographic' on Primates. Paper Tiger Television [video], 28:45. http://papertiger.org/donna-haraway-reads-the-national-geographic-on-primates/

—. 1991. Simians, Cyborgs, and Women. The Reinvention of Nature. New York: Routledge.

—. 1996. Modest Witness: Feminist Diffractions in Science Studies. P. L. Galison and D. J. Stump (eds.): The Disunity of Science: Boundaries, Contexts, and Power. Palo Alto: Stanford University Press, pages 428—442.

—. 1997. Modest_witness@second_millennium. Femaleman©_meets_oncomouse TM. Feminism and Technoscience. New York: Routledge.

—. 2013. SF: Science Fiction, Speculative Fabulation, String Figures, So Far. Ada: A Journal of Gender, New Media, and Technology 4 (November). doi:10.7264/N3KH0K81

—. 2016. Staying with the Trouble: Making Kin in the Chthulucene. Durham: Duke University Press.

Harding, Sandra 2008. Sciences from Below: Feminisms, Postcolonialities, and Modernities: Feminisms, Postcolonialisms, and Modernities. Durham: Duke University Press.

Hasse 2019. Material Concept Formation: Inequality in Children's Conceptual Robot Imaginaries. Idem and D. M. Søndergaard (eds.): Designing Robots, Designing Humans, New York: Routledge, pages 88—110.

Hayles, N. Katherine 2005. Computing the Human. Theory, Culture & Society, 22/1, pages 131—151. doi: 10.1177/0263276405048438

Hicks, Mar 2017. Programmed Inequality. Cambridge: MIT Press.

Heßler, Martina 2001. "Mrs. Modern Woman" Zur Sozial- und Kulturgeschichte der Haushaltstechnisierung. Frankfurt/Main: Campus.

Ingold, Tim 2009. The Textility of Making. Cambridge Journal of Economics, 2010/34, pages 91—102. doi:10.1093/cje/bep042

—. 2010. Bringing Things to Life: Creative Entanglements in a World of Materials, ESRC National Centre for Research Methods, NCRM Working Paper Series, 05/10, pages 2—14. http://eprints.ncrm.ac.uk/1306/1/0510_creative_entanglements.pdf

Irrgang, Melanie 2019. #5 Knitting for Computer Scientists and Engineers: Motion Planning. Do Robots Dream of Knitting?—Träumen Roboter vom Stricken? (Blog), TU Berlin, February 6 2019, https://blogs.tu-berlin.de/zifg_stricken-mit-robotern/2019/02/06/5-knitting-for-computer-scientists-and-engineers-motion-planning/

Kelly, Maura 2013. Knitting as a Feminist Project? Women's Studies International Forum, 44, pages 133—144. doi: 10.1016/j.wsif.2013.10.011

Lipp, Benjamin M. 2019. Interfacing RobotCare. On the Techno-Politics of Innovation, doctoral dissertation, Munich Center for Technology in Society, TUM. https://mediatum.ub.tum.de/doc/1472757/1472757.pdf

Leigh Star, Susan 1995. Ecologies of Knowledge: Work and Politics in Science and Technology. New York: State University of New York Press.

Lykke, Nina 2010. Feminist Studies. A Guide to Intersectional Theory, Methodology and Writing. New York: Routledge.

Lupton, Deborah 2015. Digital Sociology. New York: Routledge.

—. 2016. The Quantified Self. Cambridge and Boston: Polity.

—. 2019. Data Selves. More-than-Human Perspectives. Cambridge and Boston: Polity.

Monteiro, Stephen 2017. The Fabric of Interface. Mobile Media, Design, and Gender. Cambridge: MIT Press.

Moran, Michael E. 2007. Evolution of Robotic Arms, Robotic Surgery 1/2, pages 103-111. doi: 10.1007/s11701-006-0002-x

Mac Kenzie, Daniel and Judy Wajcman 1999. Introductory Essay. The Social Shaping of Technology, 2nd Ed. Idem (eds.): The Social Shaping of Technology. Open University Press: Buckingham, pages 3—27.

Noble, Safiya 2018. Algorithms of Oppression: How Search Engines Reinforce Racism. Combined Academic Publishers.

O'Neill, Kathy 2016. Weapons of Math Destruction: How Big Data Increases Inequality and Threatens Democracy. London: Penguin.

Pentney, Ann 2008. Knitting and Feminism's Third Wave. thirdTspace. a journal of feminist theory & culture, 8/1. https://journals.sfu.ca/thirdspace/index.php/journal/article/viewArticle/pentney/210

Pérez-Busto, Tania 2017. Thinking with Care. Unraveling and mending in an ethnography of craft embroidery and technology. Revue d'anthropologie des connaissances 11/1, pages a—u. doi 10.3917/rac.034.a

Pfeiffer, Sabine 2018. Industry 4.0: Robotics and Contradictions. P. Bilić, J. Primorac and B. Valtýsson (eds.): Technologies of Labour and the Politics of Contradiction. Camden: Palgrave Macmillan, pages 19—36. doi.org/10.1007/978-3-319-76279-1_2

Plant, Sadie 1998. Zeros+Ones: Digital Women and the New Technoculture. London: Fourth Estate.

Reiche, Claudia 2001. Die avatarische Hand. U. Bergermann, A. Sick and A. Klier (eds.), Hand. Medium¬Körper¬Technik, Bremen: thealit, pages 120—133.

Rhee, Jennifer 2018. The Robotic Imaginary. Minnesota: University of Minnesota Press.

Robben, Bernard 2012. Die Bedeutung der Körperlichkeitfür be-greifbare Interaktion mit dem Computer. Idem and H. Schelhowe (eds.): Be-greifbare InteraktionenDer allgegenwärtige Computer: Touchscreens, Wearables, Tangibles und Ubiquitous Computing, pages 19—40. Bielefeld: transcript.

Robertson, Jennifer 2010. Gendering Humanoid Robots: Robo-Sexism in Japan. Body & Society, 16/2, pages 1—36.

Rosner, Daniela K. 2018. Critical Fabulations. Cambridge: MIT Press.

Rosner, Daniela & Kimiko Ryokai 2008. Spyn: Augmenting Knitting to Support Storytelling and Reflection. UbiComp'08, September 21-24, 2008, Seoul, Korea, pages 340—349. doi: 10.1145/1409635.1409682

Schaal, Stefan 2007. The New Robotics—Towards Human-Centered Machines. HFSP Journal, 1/2, pages 115—126.

Schneider, Birgit 2007. Textiles Prozessieren. Zürich: Diaphanes.

Schrader, Astrid. 2010. Responding to Pfiesteria Piscicida (the Fish Killer): Phantomatic Ontologies, Indeterminacy, and Responsibility in Toxic Microbiology. Social Studies of Science, 40/2, pages 275–306. doi:10.1177/0306312709344902

Schwartz Cowan, Ruth 1983. More Work for Mother. New York: Basic Books.

Sinclair, Rose 2015. Dorcas Legacies, Dorcas Futures: Textile Legacies and the Formation of Identities in 'Habitus' Spaces. Craft Research, 6/2, pages 209—214. doi: 10.1386/crre.6.2.209_1

Sollfrank, Cornelia 2018. Preface. Idem (ed.): Die schönen Kriegerinnen. Technofeministische Praxis im 21. Jahrhundert, pages 7—32. Wien: transversal. https://transversal.at/media/femtec_CmBUyFV.pdf

Springgay, Stephanie 2010. Knitting as an Aesthetic of Civic Engagement: Re-conceptualizing Feminist Pedagogy through Touch. Feminist Teacher, 20/2, pages 111—123. doi: 10.5406/femteacher.20.2.0111

Suchman, Lucy 2007. Human-Machine Reconfigurations. Plans and Situated Actions, 2nd Edition. New York: Cambridge University Press.

—. 2011. Subject Objects. Feminist Theory, 12/2, pages 119 – 145.

Treusch, Pat 2015. Robotic Companionship. The Making of Anthropomatic Kitchen Robots in Queer Feminist Technoscience Perspective. Linköping: LiU Press. http://urn.kb.se/resolve?urn=urn:nbn:se:liu:diva-118117

—. 2017. The Art of Failure in Robotics: Queering the (Un)Making of Success and Failure in the Companion Robot Laboratory. B. Subramaniam and A. Willey (eds.): Science Out of Feminist Theory, Special Issue of Catalyst. Feminsim, Theory, Technoscience, 3/2, pages 1—27.

Treusch, Pat, Arne Berger and Daniela Rosner 2020. Useful Uselessness? Teaching Robots to Knit with Humans. In Proceedings of the 2020 DIS Conference on Designing Interactive Systems (DIS) 2020. ACM. New York. USA, 11 pages. doi:10.1145/3357236.3395582

van der Tuin, Iris 2018. Diffraction. R. Braidotti and M. Hlavajova (eds.) Posthuman Glossary, pages 99—101. London: Bloomsbury.

Turing, Alan. 1950. Computing Machinery and Intelligence. Mind, 59/236, pages 433–460.

Turkle, Sherry [1984] 2005. The Second Self. Computers and the Human Spirit. Twentieth Anniversary ed. Cambridge: The MIT Press.

Urry, John 2016. What is the Future? Cambridge and Boston: Polity Press.

Wallnöfer, Elsbeth 2011. Juckende Strunpfhosen und andere außerhäusliche Gemütlichkeiten. Critical Crafting Circle (eds.): Craftista! Handarbeiten als Aktivismus, pages 43-48. Mainz: Ventil Verlag.

Wajcman 2004. Technofeminism. Cambridge and Boston: Polity Press.

—. 2015. Pressed for Time: The Acceleration of Life in Digital Capitalism. Chicago: University of Chicago Press.

—. 2017. Automation: Is It Really Different This Time? The British Journal of Sociology, 68/1, pages 119—127. http://eprints.lse.ac.uk/28638/

—. 2018. Digital Technology, Work Extension and the Acceleration Society. German Journal for Ressource Management, 32/3-4, pages 168–176.

Weber, Jutta 2005. Helpless Machines and True Loving Care Givers: A Feminist Critique of Recent Trends in Human-Robot Interaction. Info, Comm & Ethics in Society, 2005/3, pages 209—218.

—. 2010. Interdisziplinierung? Zum Wissenstransfer zwischen den Geistes-, Sozial- und Technowissenschaften. Bielefeld: transcript.

—. 2014. Opacity versus Computational Reflection. Modelling Human-Robot Interaction in Personal Service Robotics. Science, Technology & Innovation Studies, 10/1, pages 187—199.

Wiescholek, Sybille 2019. Textile Bildung im Zeitalter der Digitalisierung: Vermittlungschancen zwischen Handarbeit und Technisierung. Bielefeld: transcript.

Wilson, Elizabeth A. 2010. Affect and Artificial Intelligence. University of Washington Press: Seattle.

Zweig, Anna-Katharina 2018. Wo Maschinen irren können. www.bertelsmann-stiftung.de/fileadmin/files/BSt/Publikationen/GrauePublikationen/WoMaschinenIrren Koennen.pdf, doi: 10.11586/2018006

List of Figures

Figure 1: The needle with the markers, shot by Pat Treusch
Figure 2: Screenshot of MoCap Data, shot by Jan Martin
Figure 3: Screenshot of recorded MoCap tracking, shot by Jan Martin
Figure 4: Knitting language, compiled by Hagen Verleger
Figure 5: The diamond lace chart, compiled by Hagen Verleger
Figure 6: Playing cat's cradle in the lab, screenshot of the documentary Robotic Knitting – Interdisciplinarity live!, produced by 6sept13
Figure 7: From left to right: Anne Jellinghaus, Melanie Irrgang, Philipp Graf, Raphael Deimel, Pat Treusch, PANDA, and Jan Martin, screenshot of the documentary Robotic Knitting – Interdisciplinarity live!, produced by 6sept13
Figure 8: From left to right: PANDA, Anne Jellinghaus, and Katrin M. Kämpf, shot by Pat Treusch
Figure 9: Overview of the main lab space with robots, computer working stations, and team members, shot by Katrin M. Kämpf
Figures 10 & 11: Unboxing & Assembling PANDA, shot by Pat Treusch
Figure 12: Knitting in the robotics' lab
Figures 13, 14 & 15: The three scenarios of collaborative knitting, shot by Katrin M. Kämpf
Figures 16 & 17: Realising Scenario 1, shot by Katrin M. Kämpf
Figure 18: Working on Scenario 2, shot by Katrin M. Kämpf
Figures 19 & 20: Needles and yarn, shot by Katrin M. Kämpf
Figure 21: The 3D-printed box, shot by Katrin M. Kämpf
Figure 22: The box with the needle in the gripper's grasp, shot by Katrin M. Kämpf
Figures 23 & 24: Melanie guiding PANDA, shot by Katrin M. Kämpf
Figure 25: krn on FRANKA Desk, screenshot of the FRANKAEMIKA GUI, FRANKA Desk, shot by Melanie Irrgang

Figure 26: Adjusting settings in the joint motion app, screenshot of the FRANKAEMIKA GUI, FRANKA Desk, shot by Melanie Irrgang

Figure 27: Running the trained behaviour on the virtual robot with blending of phases, screenshot of Deimel's PhastaProMP app, shot by Melanie Irrgang

Full page illustrations at the beginning and end of this book, as well as those introducing every chapter, are the artwork of Hagen Verleger.

Acknowledgments

First and foremost, I want to thank the MTIengAge lab at TUB, funded by the BMBF, the VW Foundation, funding the DRDK project, as well as the DiGiTal Programme (*DiGiTal—Digitalisierung: Gestaltung und Transformation*), funding my postdoctoral research position at TUB, financed by the BCP (*Berliner Programm zur Förderung der Chancengleichheit für Frauen in Forschung und Lehre*). It was one of the most exciting times during my research life so far to have this opportunity to bring my interests in technofeminisms, needlework, and robotics together. However, the team of the MTIengAge lab, especially Raphael Deimel, Jan Martin, Philipp Graf, Bülent Erik, and at a later stage also Arne Maibaum, as well as my team, including Melanie Irrgang, Anne Jellinghaus and Katrin M. Kämpf, made this whole experience so much more exciting. I specifically thank Katrin for documenting the project by taking pictures. Of course—not to forget, the robot PANDA number 4 in the MTIengAge lab also majorly contributed to this.

Melanie's triple qualification as a computer scientist, feminist, and avid knitter made her an especially extraordinary asset to this project. Thank you not only for your commitment to the quirkiness of robotic knitting, but also for your vital contributions on realising robotic knitting, your patience towards the more-than-once dysfunctional robot, but also with my impatience, particularly in moments of human-machine failure, as well as for your constant work on bridging disciplines through integrative everyday practices, but also the Blog series on *Knitting for Computer Scientists and Engineers*.

I also thank Romina Becker, Press Officer for Science Communication and Social Media at TUB, for her support in developing formats and helping me with any kind of science communication. Especially thrilling was the experience of shooting a short documentary on DRDK, also funded by the VW Foundation. Thus, I thank Dirk Herzog and Jan Rödger from the filming company 6sept13 for their creative ideas and precise realisation.

During the project, I had the chance to get in a conversation with a multitude of great researchers, conversations that continued after the end of DRDK. I thank Susann Fegter and her team—my colleagues at the Department of General and Historical Educational Sciences, as well as Sabine Hark and her team—my colleagues at the Center for Interdisciplinary Women's and Gender Studies (ZIFG), both at TUB. Furthermore, without being able to name them all, I would like to especially thank Daniela Rosner, Stefan Ullrich, Claude Draude, Berit Greinke, Kasper Støy, Melanie Stilz, Josef Barla, Morten Frederiksen, Cathrine Hasse, Yana Boeva, Goda Klumbyte, Loren Britton, Käthe von Bose, Mike Laufenberg, Arne Berger, Nadja Damm, Andreas Bischof and finally also Benjamin Lipp—especially for his last-minute reading of a part of this book. One of the researchers who I met through robotic knitting, Jessica Sorenson, also became a writing companion to this book. I thank you for your careful and productive editing of this book. Finally, the artwork assembled in this book is the result of my collaboration with Hagen Verleger, whom I thank for creating the images to represent robotic knitting.